Operator Theory
Advances and Applications
Vol. 97

Editor:
I. Gohberg

Editorial Office:
School of Mathematical
Sciences
Tel Aviv University
Ramat Aviv, Israel

Editorial Board:
J. Arazy (Haifa)
A. Atzmon (Tel Aviv)
J. A. Ball (Blackburg)
A. Ben-Artzi (Tel Aviv)
H. Bercovici (Bloomington)
A. Böttcher (Chemnitz)
L. de Branges (West Lafayette)
K. Clancey (Athens, USA)
L. A. Coburn (Buffalo)
K. R. Davidson (Waterloo, Ontario)
R. G. Douglas (Stony Brook)
H. Dym (Rehovot)
A. Dynin (Columbus)
P. A. Fillmore (Halifax)
C. Foias (Bloomington)
P. A. Fuhrmann (Beer Sheva)
S. Goldberg (College Park)
B. Gramsch (Mainz)
G. Heinig (Chemnitz)
J. A. Helton (La Jolla)
M.A. Kaashoek (Amsterdam)

T. Kailath (Stanford)
H.G. Kaper (Argonne)
S.T. Kuroda (Tokyo)
P. Lancaster (Calgary)
L.E. Lerer (Haifa)
E. Meister (Darmstadt)
B. Mityagin (Columbus)
V. V. Peller (Manhattan, Kansas)
J. D. Pincus (Stony Brook)
M. Rosenblum (Charlottesville)
J. Rovnyak (Charlottesville)
D. E. Sarason (Berkeley)
H. Upmeier (Marburg)
S. M. Verduyn-Lunel (Amsterdam)
D. Voiculescu (Berkeley)
H. Widom (Santa Cruz)
D. Xia (Nashville)
D. Yafaev (Rennes)

Honorary and Advisory
Editorial Board:
P. R. Halmos (Santa Clara)
T. Kato (Berkeley)
P. D. Lax (New York)
M. S. Livsic (Beer Sheva)
R. Phillips (Stanford)
B. Sz.-Nagy (Szeged)

M.G. Krein's Lectures on Entire Operators

M.L. Gorbachuk
V.I. Gorbachuk

Birkhäuser Verlag
Basel · Boston · Berlin

Authors:

M.L. Gorbachuk and
V.I. Gorbachuk
Institute of Mathematics
Ukrainian National Academy of Sciences
Tereshchenkivska Str. 3
252601 Kiev
Ukraine

1991 Mathematics Subject Classification 47B99, 47A57, 35P99

A CIP catalogue record for this book is available from the
Library of Congress, Washington D.C., USA

Deutsche Bibliothek Cataloging-in-Publication Data
Gorbačuk, Miroslav L.:
M. G. Krein's lectures on entire operators / M. L. Gorbachik ; V. I.
Gorbachuk. – Basel ; Boston ; Berlin : Birkhäuser, 1997
   (Operator theory ; Vol. 97)
   ISBN 3-7643-5704-5

© 1997 Birkhäuser Verlag, P.O. Box 133, CH-4010 Basel, Switzerland
Printed on acid-free paper produced from chlorine-free pulp. TCF ∞
Cover design: Heinz Hiltbrunner, Basel
Printed in Germany
ISBN 3-7643-5704-5
ISBN 0-8176-5704-5

Contents

Preface

This book is devoted to the theory of entire Hermitian operators, an important branch of functional analysis harmoniously combining the methods of operator theory and the theory of analytic functions. This theory anables various problems of classical and modern analysis to be looked at from a uniform point of view. In addition, it serves as a source for setting and solving many new problems in both theories. The three chapters of the book are based on the notes written by his students of M.G. Krein's lectures on the theory of entire operators with $(1,1)$-deficiency index which he delivered in 1961 at the Pedagogical Institute of Odessa, and on his works on the extension theory of Hermitian operators and the theory of analytic functions. The theory is further developed in the direction of solving the problems set up by Krein at ICM-66 in the first two appendices. The first concerns the case of Hermitian operators with arbitrary defect numbers, entire with respect to an ordinary gauge and to a generalized one as well. The other focuses on the entire operators representable by differential operators. The third appendix is the translation from Russian of the unpublished notes of Krein's lecture in which, in particular, the place of the theory of entire operators in the whole analysis is elucidated.

In Krein's mathematical heritage the theory of entire operators occupies a special position. From 1942 its intrinsic development as well as its connections with other fields of mathematics and applications were always subject of interest to him. In Krein's own reminiscences, the conceptual moments of this theory were drawn in his mind in the end of World War II. The changes in the character of the war after the Battle of Stalingrad (at the time he had been evacuated to Kuybyshev (now known as Samara)) revived his hope of coming back home to his beloved Odessa. This doubled his inspiration and by 1944 some articles had appeared in Dokl. Akad. Nauk SSSR, where the principal aspects of the theory of entire Hermitian operators whose deficiency index is $(1,1)$ were announced. This theory led him to adopt the same approach to investigate many classical problems of analysis, so different at first sight in their statements and the methods used to solve them. The power moment problem, the continuation problem for positive definite functions given on a finite interval and for spiral arcs, the description of all spectral functions of a string, the interpolation of functions and others are among them. In Krein's words, for any of these problems he saw a Hilbert space and a Hermitian operator on it hidden behind the scenes. It is to be noted that a number of new problems in the theory of analytic functions arose during the development of the theory of entire operators and its extension to more complicated situations.

In 1949 Krein developed the theory of entire operators with arbitrary finite defect numbers. This made it possible to solve the above problems in the matrix case. It should be noted that one would come across considerable analytical difficulties trying to deal with each problem separately. The case of infinite deficiency index had remained unsolved for a long time. This excited Krein because this was the case which allowed various partial differential equations problems to be considered. Krein emphasized this connection in his lectures and conversations with colleagues.

In 1966 a joint article by M. Krein and Sh. Saakyan devoted to entire operators with infinite defect numbers appeared in Dokl. Akad. Nauk SSSR. However, in the majority of problems which are simulated by differential equations, the associated operators are entire with respect to a generalized not ordinary gauge. This very fact probably motivated Krein in his one-hour lecture at the International Congress of Mathematicians (Moscow, 1966), to pay so much attention to the state of the general theory of entire operators and its relations to associated fields of mathematics, and to set the following three problems of great importance for the further development of this theory: 1) to investigate the case where a Hermitian operator is entire with respect to a gauge consisting of generalized elements; 2) to construct examples of entire operators generated by partial differential expressions; 3) to represent an arbitrary self-adjoint operator in the form of a differential one.

Unfortunately, Krein's results on entire operators have rarely been accessible for extensive study, for they are mostly announced, not published in detail. This situation troubled him greatly as the results were hardly known outside Odessa. Only his students and colleagues in Odessa were privileged to hear a detailed course of lectures on this topic. To publish it was Krein's dream. In due course the authors of this book agreed to prepare the lectures for publishing. By the end of the seventies the version adapted and extended according to Krein's suggestions had been prepared, but it was not published, because Krein became ill. The manuscript had been passed from hand to hand until in 1994 Prof. I. Gohberg proposed that we prepare it for publication. After certain modifications and additions it is presented in the form of this book. Krein's unpublished lectures on the theory of entire operators, whose deficiency index is $(1, 1)$, and his paper in Ukr. Math. J. (1949), underlie all three chapters. The main concepts of this theory and its applications are given in accordance with these lectures. The missing proofs are taken from Krein's works on the operator theory and the theory of analytic functions.

There are three appendices. Appendix 1 covers the joint results by Krein and Saakyan on Hermitian operators with arbitrary defect numbers, entire with respect to an ordinary gauge, as well as the results by Yu. Shmulian concerning the case of a generalized gauge. Appendix 2 is devoted to solving problem 2). It is proved here that many operators representable in the form of a partial differential expression and corresponding boundary condition are entire. For the sake of simplicity of the exposition we consider differential expressions with unbounded operator coefficients. This allows us to cover a number of systems (finite and infinite) of ordinary differential equations and partial differential equations of hyperbolic type. Appendix 3 consists of the notes of Krein's talk at the Jubilee session of Moscow Mathematical Society (Moscow, 1964). It was written down by one of his students. The notes were taken from Krein's archive. Perhaps they do not reflect word by word all the nuances of that talk, but they give a vibrant idea of the theory of entire operators and its place in the general operator theory and its applications, and in paticular, in Krein's own research.

It should be noted that the book does not cover many other investigations by Krein , his students and followers, concerning the theory of entire operators and its applications. First of all, we would like to mention here the description of all self-adjoint extensions of an entire Hermitian operator whose distribution functions possess certain preassigned properties (for instance, the property of the support of a spectral function belonging to a certain set). We have not touched on the applications to the interpolation and extrapolation of functions at all, though they occupied a leading position in Krein's investigations during the last years of his life. Neither have we dwell on entire operators with a nondense domain. A partial discussion of the mentioned questions can be found in the survey by E. Tzekanovsky and Yu. Shmulian [1] and in the parer by V. Derkach and M. Malamud [1]. Our main purpose was to follow the spirit of Krein's unpublished lectures, and in doing so, to pay a tribute to one of the greatest mathematicians of the centure. He believed that the theory of entire operators would contribute significantly to the further development of analysis, once more proving its inherent unity.

Invaluable help in the preparation of the book was given us by Krein's students Professors Vadim Adamyan, Damir Arov and Adolf Nudelman (Odessa), to whom we are obliged for the sketches of his lectures . Professor Anatoly Kochubei (Kiev) was greatly assisted in improving the English version of the book. The text was composed on the computer given as a gift to the Institute of Mathematics (Kiev) by the brothers Myroslav and Lubomyr Prytulak (Canada). We are greatful to all of them. Certainly, the book could never have been published without Professor Israel Gohberg's (Israel) essential support. We are also thankful to him for his advice and useful remarks.

Chapter 1

Some Aspects of Operator Theory
and the Theory of Analytic Functions

This chapter consists of four sections. Some aspects of the theory of closed Hermitian operators in a Hilbert space and their self-adjoint extensions within the given space, as well as with exit to a larger one, which are needed in what follows, are presented (without proof) in Section 1. Section 2 is devoted to the foundations of the representation theory of Hermitian operators whose deficiency index is (1,1). The main result consists of the isomorphic identification of the space where the operator acts, with a certain space of functions meromorphic inside the upper and lower half-planes. The operator itself is transformed into multiplication by the independent variable under this isomorphism. The results of Section 3 concern the structure of a spectrum of self-adjoint extensions within the original space of a simple Hermitian operator with defect numbers equal to 1. A special class of analytic functions which is of great importance for constructing the theory of entire operators is investigated in Section 4. This is the class of the so-called N-functions. The necessary and sufficient conditions for a function analytic in a disk or in the upper (lower) half-plane to be an N-function are given. The most important result of the section is the criterion for an entire function to belong to the class of N-functions in the upper (lower) half-plane. It is also established that the indicator diagram of such a function coincides with a certain interval of the imaginary axis. The asymptotic distribution of its zeroes is found on this basis.

1 On Spectral Functions of Hermitian Operators

1.1 Let $\mathfrak{H}$ be a separable Hilbert space (its completeness is always assumed), $(\cdot, \cdot)$ and $\| \cdot \|$ the scalar product and the norm in it. A linear operator A on $\mathfrak{H}$ is called closed if the existence of the limits

$$\lim_{n \to \infty} f_n = f, \quad \lim_{n \to \infty} A f_n = g, \quad f_n \in \mathcal{D}(A),$$

($\mathcal{D}(\cdot)$ is the domain of the operator) implies $f \in \mathcal{D}(A)$ and $Af = g$.

An operator $\widetilde{A}$ is called an extension of A (we write $\widetilde{A} \supset A$ or $A \subset \widetilde{A}$) if $\mathcal{D}(\widetilde{A}) \supset \mathcal{D}(A)$ and $\widetilde{A}f = Af$ for any $f \in \mathcal{D}(A)$. If the operator A itself is not closed, but if it follows from $f_n \to 0$ and $Af_n \to g$ that $g = 0$, then the operator

A is closeable, i.e. it has closed extensions. Denote by $\overline{A}$ the minimal one. It is the closure of A. To obtain $\overline{A}$ the set $\mathcal{D}(A)$ should be complemented with all vectors $f \in \mathfrak{H}$ for which there exists a sequence of elements $f_n \in \mathcal{D}(A)$ such that $f_n \to f$ and $Af_n \to g$, $n \to \infty$; then we set $\overline{A}f = g$.

The totality of all pairs $\{f, Af\}, f \in \mathcal{D}(A)$ in the direct sum $\mathfrak{H}\dotplus\mathfrak{H}$ forms the graph of A. Obviously, A is closed if and only if its graph is a closed set in $\mathfrak{H}\dotplus\mathfrak{H}$.

The following statements hold:

(i) if A is closed and B ($\mathcal{D}(B) = \mathfrak{H}$) is a bounded operator, then the operator $A + B$ is closed;

(ii) if A is closed and A^{-1} exists, then the operator A^{-1} is closed;

(iii) the closed graph theorem: if a closed operator is given on the whole space, then it is bounded.

Define on $\mathcal{D}(A)$ the scalar product

$$(f, g)_A = (f, g) + (Af, Ag), \quad f, g \in \mathcal{D}(A). \tag{1.1}$$

If A is closed, then $\mathcal{D}(A)$ is a Hilbert space with respect to the scalar product (1.1).

Suppose the domain $\mathcal{D}(A)$ to be dense in $\mathfrak{H}$ (this is denoted by $\overline{\mathcal{D}(A)} = \mathfrak{H}$). In what follows we consider only closed linear operators with domains dense in $\mathfrak{H}$.

Let $\mathbb{C}^1$ denote the complex plane. A number $z \in \mathbb{C}^1$ is called a regular point of the operator A if the operator

$$R_z(A) = (A - zI)^{-1}$$

(I is the identity operator) exists as a bounded operator with the whole space $\mathfrak{H}$ as its domain. We denote by $\rho(A)$ the set of all regular points of A. The operator-function $R_z(A)$ defined on $\rho(A)$ is called the resolvent of A. The set $\rho(A)$ is always open and the resolvent $R_z(A)$ is analytic on it. For any $z, \zeta \in \rho(A)$ the Hilbert resolvent identity

$$R_z(A) - R_\zeta(A) = (z - \zeta)R_z(A)R_\zeta(A) \tag{1.2}$$

holds. Note also that $R_z(A)R_\zeta(A) = R_\zeta(A)R_z(A)$.

The complement of $\rho(A)$ to $\mathbb{C}^1$ is called the spectrum of the operator A. We denote it by $\sigma(A)$. Consequently, the set $\sigma(A)$ is always closed. We use the following classification of points in the spectrum.

The set $\sigma_p(A)$ of numbers $z \in \sigma(A)$ such that the map $A - zI$ is not one-to-one is called the point spectrum of A. Thus, $z \in \sigma_p(A)$ if and only if $Af = zf$ for some $f \neq 0$, i.e. z is an eigenvalue of the operator A and f is the eigenvector of A, corresponding to z. The set of all eigenvectors of A corresponding to z forms a subspace in $\mathfrak{H}$. Its dimension is called the multiplicity of the eigenvalue z.

The set $\sigma_c(A)$ of all $z \in \sigma(A)$ such that the map $A - zI$ is one-to-one and $\overline{(A - zI)\mathcal{D}(A)} = \mathfrak{H}$ but $(A - zI)\mathcal{D}(A) \neq \mathfrak{H}$ is called the continuous spectrum of A.

The set $\sigma_r(A)$ of all $z \in \sigma(A)$ for which $A - zI$ is one-to-one and $(A - zI)\mathcal{D}(A)$ is not dense in $\mathfrak{H}$ is called the residual spectrum of A. Clearly, $\sigma_p(A), \sigma_c(A)$ and $\sigma_r(A)$ are disjoint and $\sigma(A) = \sigma_p(A) \cup \sigma_c(A) \cup \sigma_r(A)$.

Let the elements $g, g^* \in \mathfrak{H}$ be such that

$$(Af, g) = (f, g^*) \tag{1.3}$$

for any $f \in \mathcal{D}(A)$. In view of the density of $\mathcal{D}(A)$ in $\mathfrak{H}$, the element g^* is uniquely determined by g. So, the correspondence $g \to g^*$ defines the linear operator A^*, called the adjoint of A. The set of all the elements g for which there exists g^* such that (1.3) holds, forms $\mathcal{D}(A^*)$.

The operator A is called Hermitian if $A \subseteq A^*$, that is $\mathcal{D}(A) \subseteq \mathcal{D}(A^*)$ and

$$(Af, g) = (f, Ag)$$

for arbitrary $f, g \in \mathcal{D}(A)$.

If $A = A^*$, then the operator A is called self-adjoint. In this case $\sigma(A) \subset \mathbb{R}^1$ and $\sigma_r(A) = \emptyset$.

1.2 Let A $(\overline{\mathcal{D}(A)} = \mathfrak{H})$ be a closed Hermitian operator on $\mathfrak{H}$, and $z \in \mathbb{C}^1, \Im z \neq 0$. Put

$$\mathfrak{M}_z = (A - zI)\mathcal{D}(A).$$

The set $\mathfrak{M}_z$ is a subspace of $\mathfrak{H}$. Its orthogonal complement

$$\mathfrak{N}_{\bar{z}} = \mathfrak{H} \ominus \mathfrak{M}_z$$

to $\mathfrak{H}$ coincides with the eigensubspace corresponding to the eigenvalue $\bar{z}$ of the operator A^*. Indeed, if $g \in \mathfrak{N}_{\bar{z}}$, then

$$(Af - zf, g) = 0 \tag{1.4}$$

for every vector $f \in \mathcal{D}(A)$, whence

$$(Af, g) = (f, \bar{z}g). \tag{1.5}$$

By the definition of the adjoint of an operator , $g \in \mathcal{D}(A^*)$ and $A^*g = \bar{z}g$. Conversely, the equality $A^*g = \bar{z}g$ implies (1.5), which is equivalent to (1.4).

We shall call the linear sets $M_1, \ldots, M_n$ from $\mathfrak{H}$ linearly independent if the equality

$$f_1 + f_2 + \ldots + f_n = 0, \quad f_k \in M_k,$$

is possible only for $f_k = 0$, $k = 1, \ldots, n$.

The sets $\mathcal{D}(A)$, $\mathfrak{N}_z$ and $\mathfrak{N}_{\bar{z}}$ $(\Im z \neq 0)$ are linearly independent,

$$\mathcal{D}(A^*) = \mathcal{D}(A) \dot{+} \mathfrak{N}_z \dot{+} \mathfrak{N}_{\bar{z}} \tag{1.6}$$

and

$$A^*f = Af_0 + zf_z + \bar{z}f_{\bar{z}} \tag{1.7}$$

for any

$$f = f_0 + f_z + f_{\bar{z}}$$

where $f_0 \in \mathcal{D}(A), f_z \in \mathfrak{N}_z$. The numbers $n_+ = \dim \mathfrak{N}_z$ and $n_- = \dim \mathfrak{N}_{\bar{z}}$ do not change when z runs through the half-plane $\Im z > 0$. The pair (n_+, n_-) is called the deficiency index of the operator A, while the numbers n_+ and n_- are called its defect numbers. The formulas (1.6),(1.7) show that a Hermitian operator is self-adjoint if and only if $n_+ = n_- = 0$.

Let $\mathfrak{M}$ and $\mathfrak{N}$ be subspaces in $\mathfrak{H}$. The number n is called the dimension of $\mathfrak{N}$ modulo $\mathfrak{M}$ (denoted by $\dim_{\mathfrak{M}} \mathfrak{N}$) if there exist n, and not more than n vectors in $\mathfrak{N}$, no linear combination of which except for the combination with zero coefficients belongs to $\mathfrak{M}$. It follows from (1.6) that

$$\dim_{\mathcal{D}(A)} \mathcal{D}(A^*) = n_+ + n_-.$$

A closed Hermitian operator admits self-adjoint extensions within the original space $\mathfrak{H}$ if and only if its defect numbers are equal. The formulas

$$\mathcal{D}(\widetilde{A}) = \mathcal{D}(A) + (U + I)\,\mathfrak{N}_{\bar{z}} \tag{1.8}$$

and

$$\widetilde{A}f = f_0 + zU f_{\bar{z}} + \bar{z} f_{\bar{z}}, \tag{1.9}$$

where U is a unitary operator from $\mathfrak{N}_{\bar{z}}$ onto $\mathfrak{N}_z$, give a one-to-one correspondence between the set of all self-adjoint extensions $\widetilde{A}$ within $\mathfrak{H}$ of A and the set of all unitary operators U acting from $\mathfrak{N}_{\bar{z}}$ onto $\mathfrak{N}_z$. The relations (1.8) and (1.9) show that the condition

$$\dim_{\mathcal{D}(A)} \mathcal{D}(\widetilde{A}) = n_+ = n_-$$

is necessary and sufficient for the operator $\widetilde{A} : A \subset \widetilde{A} \subset \widetilde{A}^*$ to be a self-adjoint extension within $\mathfrak{H}$ of A.

It turns out that if the defect numbers of the operator A are equal and finite, then the continuous spectrum of all its self-adjoint extensions within $\mathfrak{H}$ is the same.

In the case of $n_+ \neq n_-$ the operator A has no self-adjoint extension within $\mathfrak{H}$. However, there always exist a Hilbert space $\widetilde{\mathfrak{H}} \supset \mathfrak{H}$ and a self-adjoint operator $\widetilde{A}$ on $\widetilde{\mathfrak{H}}$, which is an extension of the operator A. We shall call such an operator $\widetilde{A}$ an extension with exit to the larger space $\widetilde{\mathfrak{H}}$.

1.3. We define a resolution of the identity as a one-parameter family of bounded self-adjoint operators E_λ, $\lambda \in \mathbb{R}^1$ on $\mathfrak{H}$ satisfying for any $f \in \mathfrak{H}$ the following conditions:

1) the function $(E_\lambda f, f)$ does not decrease when λ increases;

2) the vector-function $E_\lambda f$ is left-continuous;

3) $E_\lambda f \to 0$ as $\lambda \to -\infty$;

4) $E_\lambda f \to f$ when $\lambda \to \infty$.

A resolution of the identity is called orthogonal if

$$E_\lambda E_\mu = E_{min\{\lambda,\mu\}}$$

for all $\lambda, \mu \in \mathbb{R}^1$.

We use the following notation. If Δ is one of the intervals (a, b), $[a, b)$, $(a, b]$, or $[a, b]$, then E_Δ denotes, respectively, $E_b - E_{a+0}$, $E_b - E_a$, $E_{b+0} - E_{a+0}$, $E_{b+0} - E_a$. In particular, $E_\lambda = E_\Delta$ for $\Delta = (-\infty, \lambda)$.

If E_λ is orthogonal, then E_Δ is an orthoprojector, $E_{\Delta_1} E_{\Delta_2} = 0$ as $\Delta_1 \cap \Delta_2 = \emptyset$, and the set $\{E_\Delta h\}$, where the finite interval Δ and an element $f \in \mathfrak{H}$ are arbitrary, is dense in $\mathfrak{H}$.

Now let E_λ be an orthogonal resolution of the identity and $f(\lambda) \in C[a, b]$, $-\infty < a < b < \infty$ ($C(M)$ denotes the set of all functions continuous on M). Take a partition of $[a, b)$ by points $\lambda_0 = a, \lambda_1, \dots, \lambda_{n-1}, \lambda_n = b$, set $\Delta_j = [\lambda_{j-1}, \lambda_j)$, and form the sum

$$S_\Delta = \sum_{j=1}^{n} f(\xi_j) E_{\Delta_j},$$

where ξ_j is an arbitrary point in the interval Δ_j. The operator S_Δ is bounded on $\mathfrak{H}$. One can show that the sum S_Δ tends to a certain operator in the uniform operator topology as $\max |\Delta_j| = \max |\lambda_j - \lambda_{j-1}| \to 0$. It does not depend on the choice of a partition of $[a, b)$ or on the points ξ_i. The integral

$$\int_a^b f(\lambda) \, dE_\lambda \qquad (1.10)$$

is understood to be this limit.

The existence of the integral (1.10) implies that of the integral

$$\int_a^b f(\lambda) \, dE_\lambda h = \lim_{\max |\Delta_j| \to 0} \sum_{j=1}^{n} f(\xi_j) E_{\Delta_j} h = \left(\int_a^b f(\lambda) \, dE_\lambda \right) h, \quad h \in \mathfrak{H}. \quad (1.11)$$

Moreover,

$$\left\| \int_a^b f(\lambda) \, dE_\lambda h \right\|^2 = \int_a^b |f(\lambda)|^2 \, d(E_\lambda h, h). \qquad (1.12)$$

On the right-hand side of this equality we have an ordinary Stieltjes integral.

If $f(\lambda) \in C(\mathbb{R}^1)$, then we can define the integral

$$\int_{-\infty}^{\infty} f(\lambda) dE_\lambda h$$

as the limit of the integral (1.11) when $a \to -\infty$, $b \to \infty$. Because of the Cauchy convergence criterion and formula (1.12), this integral exists if and only if the ordinary Stieltjes integral

$$\int_{-\infty}^{\infty} |f(\lambda)|^2 \, d(E_\lambda h, h)$$

exists, and

$$\left\| \int_{-\infty}^{\infty} f(\lambda) \, dE_\lambda h \right\|^2 = \int_{-\infty}^{\infty} |f(\lambda)|^2 \, d(E_\lambda h, h).$$

For an arbitrary self-adjoint operator A on $\mathfrak{H}$ the following statement, which is called the principal spectral theorem, is valid.

Theorem 1.1 *Let A be a self-adjoint operator on $\mathfrak{H}$. Then there exists a unique orthogonal resolution of the identity E_λ, called the spectral function of A, such that:*

(i) *a vector h belongs to $\mathcal{D}(A)$ if and only if*

$$\int_{-\infty}^{\infty} |\lambda|^2 \, d(E_\lambda h, h) \ < \infty;$$

(ii) *for any $h \in \mathcal{D}(A)$*

$$Ah \ = \ \int_{-\infty}^{\infty} \lambda \, dE_\lambda h, \quad \text{hence,} \quad \|Ah\|^2 \ = \ \int_{-\infty}^{\infty} \lambda^2 \, d(E_\lambda h, h).$$

Conversely, any operator A determined by the conditions (i) and (ii), where E_λ is an orthogonal resolution of the identity, is self-adjoint.

It is possible to characterize the spectrum of a self-adjoint operator A by means of its spectral function as follows:

a) a real number λ_0 is a regular point of the operator A if and only if its spectral function E_λ is constant in some neighbourhood of the point λ_0;

b) a real number λ_0 is an eigenvalue of A if and only if $E_{\lambda_0+0} - E_{\lambda_0} \neq 0$; in this case $E_{\lambda_0+0} - E_{\lambda_0}$ is the orthoprojector onto the eigensubspace of A corresponding to λ_0.

1.4 According to the Naimark theorem , for every resolution of the identity there exist a Hilbert space $\widetilde{\mathfrak{H}} \supset \mathfrak{H}$ and an orthogonal one $\widetilde{E}_\lambda$ on it such that

$$E_\lambda = P\widetilde{E}_\lambda, \quad \lambda \in \mathbb{R}^1, \tag{1.13}$$

where P is the orthoprojector from $\widetilde{\mathfrak{H}}$ onto $\mathfrak{H}$. It should be noted that $\widetilde{\mathfrak{H}}$ and $\widetilde{E}_\lambda$ may be chosen so that the representation (1.13) is irreducible, i.e. there is no subspace in $\widetilde{\mathfrak{H}} \ominus \mathfrak{H}$ invariant for all the operators $\widetilde{E}_\lambda$. The irreducible representation (1.13) is unique up to a unitary equivalence. This means that for any other representation

$$E_\lambda = P'\widetilde{E}'_\lambda, \quad \lambda \in \mathbb{R}^1,$$

where P' is the orthogonal projector onto $\mathfrak{H}$ from the space $\widetilde{\mathfrak{H}}' \supset \mathfrak{H}$, in which the orthogonal resolution of the identity $\widetilde{E}'_\lambda$ acts, there exists a unitary operator U from $\widetilde{\mathfrak{H}}$ onto $\widetilde{\mathfrak{H}}'$ such that $\widetilde{E}_\lambda = U\widetilde{E}'_\lambda U^{-1}$ and $U\mathfrak{H} = \mathfrak{H}$.

It follows from the representation (1.13) that if

$$\int_{-\infty}^{\infty} \lambda^2 \, d(E_\lambda f, f) < \infty$$

for some vector $f \in \mathfrak{H}$, then the integrals in the equality

$$\int_{-\infty}^{\infty} \lambda \, dE_\lambda f = P \int_{-\infty}^{\infty} \lambda \, d\widetilde{E}_\lambda f \tag{1.14}$$

converge. Indeed, in view of (1.13),

$$\int_{-\infty}^{\infty} \lambda^2 \, d(\widetilde{E}_\lambda f, f) = \int_{-\infty}^{\infty} \lambda^2 \, d(E_\lambda f, f) < \infty.$$

Since the resolution of the identity $\widetilde{E}_\lambda$ is orthogonal, the convergence of this integral implies, by Theorem 1.1, convergence of the integral in the right-hand side of (1.14).

Suppose $A \, (\overline{\mathcal{D}(A)} = \mathfrak{H})$ to be a closed Hermitian operator in $\mathfrak{H}$. A resolution of the identity E_λ is called a spectral function of the operator A if

$$\|Af\|^2 = \int_{-\infty}^{\infty} \lambda^2 \, d(E_\lambda f, f) \quad \text{and} \quad Af = \int_{-\infty}^{\infty} \lambda \, dE_\lambda f \tag{1.15}$$

for every $f \in \mathcal{D}(A)$.

Let E_λ be a spectral function of the operator A. Then $E_\lambda = P\widetilde{E}_\lambda$ where, by the principal spectral theorem, $\widetilde{E}_\lambda$ is the spectral function of a certain self-adjoint operator $\widetilde{A}$ on the space $\widetilde{\mathfrak{H}} \supset \mathfrak{H}$. Its domain $\mathcal{D}(\widetilde{A})$ consists of all the elements $f \in \widetilde{\mathfrak{H}}$ for which

$$\int\limits_{-\infty}^{\infty} \lambda^2 \, d(\widetilde{E}_\lambda f, f) < \infty, \quad \text{and if } f \in \mathcal{D}(\widetilde{A}), \text{ then} \quad \widetilde{A}f = \int\limits_{-\infty}^{\infty} \lambda \, d\widetilde{E}_\lambda f.$$

It follows from here that $\mathcal{D}(A) \subset \mathcal{D}(\widetilde{A})$. In addition, by virtue of (1.15),

$$\|Af\|^2 = \int\limits_{-\infty}^{\infty} \lambda^2 \, d(E_\lambda f, f) = \int\limits_{-\infty}^{\infty} \lambda^2 \, d(\widetilde{E}_\lambda f, f) = \|\widetilde{A}f\|^2$$

and

$$Af = \int\limits_{-\infty}^{\infty} \lambda \, dE_\lambda f = P \int\limits_{-\infty}^{\infty} \lambda \, d\widetilde{E}_\lambda f = P\widetilde{A}f$$

for any $f \in \mathcal{D}(A)$. These two equalities show that $Af = \widetilde{A}f$, $f \in \mathcal{D}(A)$, i.e. $\widetilde{A}$ is a self-adjoint extension with exit to the space $\widetilde{\mathfrak{H}}$ of the operator A.

It is not difficult to make sure that the converse statement is valid too. Namely, if $\widetilde{A}$ is a self-adjoint extension of the operator A to $\widetilde{\mathfrak{H}} \supset \mathfrak{H}$ and $\widetilde{E}_\lambda$ its spectral function, then the equality

$$E_\lambda = P\widetilde{E}_\lambda$$

determines a certain spectral function E_λ of the operator A. This representation is irreducible if and only if the extension $\widetilde{A}$ is irreducible, that is, there is no subspace of $\widetilde{\mathfrak{H}} \ominus \mathfrak{H}$, invariant for the operator $\widetilde{A}$, in which this operator is self-adjoint. As the operator A always has self-adjoint extensions with exit to a larger space, there is at least one spectral function of A. Thus we have described above the way to get all spectral functions of a closed Hermitian operator. The spectral function is unique if and only if the operator under consideration is maximal, i.e. its deficiency index is $(0, n_-)$ or $(n_+, 0)$.

2 Representation of a Simple Hermitian Operator whose Deficiency Index is (1,1)

2.1   Let A be a closed Hermitian operator with dense domain in $\mathfrak{H}$ and $\widetilde{A}$ its self-adjoint extension to $\widetilde{\mathfrak{H}}$ (in special cases the space $\widetilde{\mathfrak{H}}$ may coincide with $\mathfrak{H}$). For arbitrary $z : \Im z \neq 0$ and $\zeta : \Im \zeta \neq 0$ we put

$$\widetilde{U}_{\zeta z} = I + (z - \zeta)\widetilde{R}_z = (\widetilde{A} - \zeta I)(\widetilde{A} - zI)^{-1} \tag{2.1}$$

where $\widetilde{R}_z = R_z(\widetilde{A})$ is the resolvent of the operator $\widetilde{A}$. The operator $\widetilde{U}_{\bar{z}z}$ is the well-known Caley transform.

The operator $\widetilde{U}_{\zeta z}$ possesses the following properties:

$1°$ $\widetilde{U}_{\zeta z} = \widetilde{U}_{z\zeta}^{-1};$

$2°$ $\widetilde{U}_{z\zeta}^{*} = \widetilde{U}_{\bar{z}\bar{\zeta}};$

$3°$ $\widetilde{U}_{\zeta z}\widetilde{U}_{z\xi} = \widetilde{U}_{\zeta\xi};$

$4°$ $\widetilde{U}_{\zeta z}$ is a one-to-one map from $\widetilde{\mathfrak{N}}_{\zeta}$ onto $\widetilde{\mathfrak{N}}_{z}$, where

$$\widetilde{\mathfrak{N}}_{z} = \mathfrak{H} \ominus \mathfrak{M}_{\bar{z}}.$$

The equalities $1°$–$3°$ are easily verifiable. We therefore prove only property $4°$.

Let $\varphi \in \widetilde{\mathfrak{N}}_{\zeta}$. Then φ is orthogonal to $\mathfrak{M}_{\bar{\zeta}}$ and

$$\left(\widetilde{U}_{\zeta z}\varphi, (A - \bar{z}I)f\right) = \left(\varphi, \widetilde{U}_{\bar{\zeta}\bar{z}}(\tilde{A} - \bar{z}I)f\right) = \left(\varphi, (\tilde{A} - \bar{\zeta}I)f\right) = 0$$

for any $f \in \mathcal{D}(A)$. Consequently, $\widetilde{U}_{\zeta z}\widetilde{\mathfrak{N}}_{\zeta} \subseteq \widetilde{\mathfrak{N}}_{z}$. In the same way we obtain $\widetilde{U}_{z\zeta}\widetilde{\mathfrak{N}}_{z} \subseteq \widetilde{\mathfrak{N}}_{\zeta}$. It is clear now, by virtue of property $1°$, that assertion $4°$ is valid.

Theorem 2.1 *The set*

$$\mathfrak{H}_0 = \bigcap_{z:\Im z \neq 0} \mathfrak{M}_z$$

is the maximal subspace invariant for the operator A on which the operator A is self-adjoint.

Proof. Let $\mathfrak{L}$ be a subspace invariant for the operator A. This means that $A\mathcal{D}_1 \subset \mathfrak{L}$, where $\mathcal{D}_1 = \mathcal{D}(A) \cap \mathfrak{L}$, and the operator A is self-adjoint on $\mathfrak{L}$. So,

$$\mathfrak{L} = (A - zI)\mathcal{D}_1 \subset \mathfrak{M}_z$$

for any $z : \Im z \neq 0$. Thus, $\mathfrak{L} \subset \mathfrak{H}_0$. It only remains, therefore, to verify that the space $\mathfrak{H}_0$ is invariant for the operator A (i.e. $A(\mathcal{D}(A) \cap \mathfrak{H}_0) \subset \mathfrak{H}_0$), and the operator A is self-adjoint on $\mathfrak{H}_0$. But if $f \in \mathfrak{H}_0$, then f is orthogonal to all the subspaces $\widetilde{\mathfrak{N}}_z$, $\Im z \neq 0$. According to property $4°$ of the map $\widetilde{U}_{\zeta z}$,

$$\psi = \widetilde{U}_{\zeta z}\varphi = \varphi + (z - \zeta)\widetilde{R}_z\varphi \in \widetilde{\mathfrak{N}}_z$$

for any $\varphi \in \widetilde{\mathfrak{N}}_{\zeta}$. Hence, the orthogonality of f to $\widetilde{\mathfrak{N}}_{\zeta}$ and $\widetilde{\mathfrak{N}}_z$ implies that of f to $\widetilde{R}_z\varphi$, whence $(\widetilde{R}_{\bar{z}}f, \varphi) = 0$, i.e. $\widetilde{R}_{\bar{z}}f$ is orthogonal to $\widetilde{\mathfrak{N}}_{\zeta}$. Since $z : \Im z \neq 0$ and $\zeta : \Im \zeta \neq 0$ are arbitrary, we have $\widetilde{R}_z f \in \mathfrak{H}_0$. Thus, $\widetilde{R}_z\mathfrak{H}_0 \subset \mathfrak{H}_0$ for every $z : \Im z \neq 0$.

On the other hand, if $f \in \mathfrak{H}_0$, then $f = (A - zI)g$, $g \in \mathcal{D}(A)$, and $\widetilde{R}_z f = g \in \mathcal{D}(A)$. So, for any $z : \Im z \neq 0$,

$$\widetilde{R}_z\mathfrak{H}_0 \subset \mathcal{D}(A) \cap \mathfrak{H}_0. \tag{2.2}$$

Now let g be an element from $\mathcal{D}(A) \cap \mathfrak{H}_0$ and $f = (A - zI)g$. Then $\widetilde{R}_\zeta f = g + (\zeta - z)\widetilde{R}_\zeta g \in \mathcal{D}(A) \cap \mathfrak{H}_0$. Consequently, $f = (A - \zeta I)\widetilde{R}_\zeta f \in \mathfrak{M}_\zeta$, $\Im\zeta \neq 0$, which implies $f \in \mathfrak{H}_0$. It follows that also $Ag \in \mathfrak{H}_0$. Thus, the subspace $\mathfrak{H}_0$ is invariant for the operator A.

The foregoing arguments and the inclusion (2.2) show that

$$(A - zI)(\mathcal{D}(A) \cap \mathfrak{H}_0) = \mathfrak{H}_0,$$

which is equivalent to self-adjointness of A on $\mathfrak{H}_0$. $\square$

Definition 2.1 *A Hermitian operator A is called simple if it has no nontrivial sub-space $\mathfrak{L}$ invariant for A, such that $\overline{\mathfrak{L} \cap \mathcal{D}(A)} = \mathfrak{L}$ and the restriction of A to $\mathfrak{L} \cap \mathcal{D}(A)$ is self-adjoint on $\mathfrak{L}$.*

We can see that Theorem 2.1 gives the condition necessary and sufficient for the operator A to be simple.

Denote by $\mathfrak{H}_1$ the space $\mathfrak{H} \ominus \mathfrak{H}_0$. As the deficiency index of the restriction of A to $\mathfrak{H}_1 \cap \mathcal{D}(A)$ is equal to that of the operator A, we may assume without loss of generality that the operator A under consideration is a simple Hermitian operator, that is,

$$\bigcap_{z:\Im z \neq 0} \mathfrak{M}_z = 0.$$

In what follows we suppose that this condition is satisfied.

2.2   Let A be a simple Hermitian operator with dense domain on $\mathfrak{H}$, whose de-ficiency index is (1,1). This means that $\dim \mathfrak{N}_z = 1$ for every $z : \Im z \neq 0$. Suppose $\overset{\circ}{A}$ to be a self-adjoint extension of A within $\mathfrak{H}$. Fix a point $z_0 : \Im z_0 \neq 0$ and an element $\varphi_0 \neq 0$ from $\mathfrak{N}_{z_0}$ and define $\varphi(z)$ as

$$\varphi(z) = \overset{\circ}{U}_{z_0 z} \varphi_0 = \varphi_0 + (z - z_0)\overset{\circ}{R}_z \varphi_0, \tag{2.3}$$

where $\overset{\circ}{R}_z = R_z(\overset{\circ}{A})$.

The vector-function $\varphi(z)$ taking its values in the subspace $\mathfrak{N}(z)$ is analytic in the upper and lower half-planes and generally in every domain that consists of regular points of the operator $\overset{\circ}{A}$. Because of property 3° of the map (2.1),

$$\overset{\circ}{U}_{z_0 z} = \overset{\circ}{U}_{z_0 \zeta} \overset{\circ}{U}_{\zeta z} = \overset{\circ}{U}_{\zeta z} \overset{\circ}{U}_{z_0 \zeta},$$

and we can assert that

$$\varphi(z) = \overset{\circ}{U}_{\zeta z} \varphi(\zeta) = \varphi(\zeta) + (z - \zeta)\overset{\circ}{R}_z \varphi(\zeta)$$

for arbitrary regular points z, ζ of the operator $\overset{\circ}{A}$.

We will show that the operator A generates a certain representation of the space $\mathfrak{H}$ in which every vector $f \in \mathfrak{H}$ is associated with some function meromorphic inside the upper and lower half-planes. To this end we take a one-dimensional subspace $M \subset \mathfrak{H}$ (the so-called module of the representation) such that

$$M \cap \mathfrak{M}_z = 0 \qquad (2.4)$$

for at least one $z = z_+ : \Im z_+ > 0$ and one $z = z_- : \Im z_- < 0$. The condition (2.4) means that there is no non-zero vector in M orthogonal to $\mathfrak{N}_z$. If $u \neq 0$ is an element from M (we will call this element the gauge of the representation), then condition (2.4) is satisfied if and only if

$$(u, \varphi(\bar{z})) \neq 0.$$

As is easily seen, the module M may be chosen in the linear span of the vectors $\varphi(\bar{z}_+)$ and $\varphi(\bar{z}_-)$. Indeed, since the subspaces $\mathfrak{N}_{\bar{z}_+}$ and $\mathfrak{N}_{\bar{z}_-}$ are linearly independent, we can select the constants c_1 and c_2 so that the vector $c_1\varphi(\bar{z}_+) + c_2\varphi(\bar{z}_-)$ is not orthogonal to $\varphi(\bar{z}_+)$ and $\varphi(\bar{z}_-)$.

Denote by S_M the countable set of all the points $z : \Im z \neq 0$ where the function $(u, \varphi(\bar{z}))$ vanishes. Obviously, condition (2.4) is satisfied for $z : \Im z \neq 0$ if and only if $z \notin S_M$. If this is the case, then the space $\mathfrak{H}$ may be decomposed into the direct sum of the spaces $\mathfrak{M}_z$ and M:

$$\mathfrak{H} = \mathfrak{M}_z \dotplus M.$$

Assign to every $f \in \mathfrak{H}$ its component $f_M(z)$ in M. This component is determined by the following two conditions:

1) $f_M(z) = f_u(z)u \in M$;

2) $f - f_M(z) \in \mathfrak{M}_z$.

It follows from condition 2) that

$$\left(f - f_u(z)u,\, \varphi(\bar{z})\right) \;=\; 0,$$

whence

$$f_u(z) = \frac{(f, \varphi(\bar{z}))}{(u, \varphi(\bar{z}))}. \qquad (2.5)$$

The equality (2.5) shows that for every $f \in \mathfrak{H}$ the function $f_M(z)$ is meromorphic in each of the half-planes $\Im z > 0$ and $\Im z < 0$ and its poles belong to S_M.

Denote by $\mathfrak{H}_M$ the linear set of the functions $f_M(z)$, $\Im z \neq 0$ corresponding to all vectors $f \in \mathfrak{H}$. It is evident that if $f_M(z) \equiv 0$, then $f \in \bigcap_{z:\Im z \neq 0} \mathfrak{M}_z$, hence $f = 0$. So the linear mapping $f \mapsto f_M(z)$ is one-to-one. It remains to add a little to verify the following assertion.

Theorem 2.2 *The map*

$$f \mapsto f_M(z) \tag{2.6}$$

is a linear isomorphism from $\mathfrak{H}$ onto $\mathfrak{H}_M$. In this isomorphism the given operator A is transformed into the operator of multiplication by z. The transform (2.6) leaves the set M fixed, i.e. $f_M(z) \equiv f$ if $f \in M$.

Proof. Indeed, the last assertion is evident. It remains therefore to prove that if $f \in \mathcal{D}(A)$ and $g = Af$, then $g_M(z) = zf_M(z)$. But if $g = Af$, then $h = g - zf = (A - zI)f \in \mathfrak{M}_z$, hence $h_M(z) = g_M(z) - zf_M(z) = 0$, which completes the proof. $\qquad\square$

The set $\mathfrak{H}_M$ possesses the following property.

Theorem 2.3 *If $f_M(z) \in \mathfrak{H}_M$, then*

$$\frac{f_M(z) - f_M(a)}{z - a} \in \mathfrak{H}_M, \tag{2.7}$$

where a is any non-real number which is not a pole of the function $f_M(z)$.

Proof. Let f be the element associated with $f_M(z)$. Set $h = f - f_M(a)$. Then $h_M(z) = f_M(z) - f_M(a)$, because the mapping (2.6) leaves the elements from M fixed. In particular, $h_M(a) = 0$, whence $h \in \mathfrak{M}_a$. Denote by g the element from $\mathcal{D}(A)$ such that $(A - aI)g = h$. Then $(z - a)g_M(z) = h_M(z)$. Hence the function (2.7) coincides with $g_M(z)$. $\qquad\square$

We need the following important statement.

Theorem 2.4 *Let $\tau(\lambda) = \frac{1}{2}[\tau(\lambda + 0) + \tau(\lambda - 0)]$, $\lambda \in \mathbb{R}^1$, be a function of bounded variation in every finite interval, such that the integral*

$$F(z) = \int\limits_{-\infty}^{\infty} \frac{d\tau(\lambda)}{\lambda - z}, \quad \Im z \neq 0 \tag{2.8}$$

converges absolutely. If $\varphi(\lambda)$ is an analytic function on a closed interval $[a, b]$, then

$$\lim_{\varepsilon \to 0} \frac{1}{2\pi i} \int\limits_{\Delta_\varepsilon} \varphi(z) F(z)\, dz = -\int\limits_{a}^{b} \varphi(\lambda)\, d\tau(\lambda). \tag{2.9}$$

Here the broken path $\Delta_\varepsilon, \varepsilon > 0$, consists of the directed intervals $(a - i\varepsilon, b - i\varepsilon)$ and $(b + i\varepsilon, a + i\varepsilon)$.

This theorem is an immediate generalization of the well-known Stieltjes inversion rule for the integral (2.8). We will obtain this rule as a special case of the theorem if we take $\varphi(\lambda) \equiv 1$.

Proof. Suppose first that a and b are points of continuity for the function $\tau(\lambda)$, and set

$$F_1(z) = \int_a^b \frac{d\tau(\lambda)}{\lambda - z}, \qquad F_2(z) = \left(\int_{-\infty}^a + \int_b^\infty \right) \frac{d\tau(\lambda)}{\lambda - z}.$$

The theorem will be proved if we show that

$$\lim_{\varepsilon \to 0} \frac{1}{2\pi i} \int_{\Delta_\varepsilon} \varphi(z) F_1(z)\, dz = - \int_a^b \varphi(\lambda)\, d\tau(\lambda) \tag{2.10}$$

and

$$\lim_{\varepsilon \to 0} \frac{1}{2\pi i} \int_{\Delta_\varepsilon} \varphi(z) F_2(z)\, dz = 0. \tag{2.11}$$

Denote by Γ_ε the path composed of the intervals $(a - i\varepsilon, b - i\varepsilon), (b + i\varepsilon, a + i\varepsilon)$ and the semicircles

$$K_{1\varepsilon} = \{ z : z - a = \varepsilon e^{i\varphi}, \ \frac{\pi}{2} \le \varphi \le \frac{3\pi}{2} \}$$

and

$$K_{2\varepsilon} = \{ z : z - b = \varepsilon e^{i\varphi}, \ -\frac{\pi}{2} \le \varphi \le \frac{\pi}{2} \}.$$

Since the contour Γ_ε envelops the interval $[a, b]$,

$$\frac{1}{2\pi i} \int_{\Gamma_\varepsilon} \varphi(z) F_1(z)\, dz$$

$$= \int_a^b \left(\frac{1}{2\pi i} \int_{\Gamma_\varepsilon} \frac{\varphi(z)}{\lambda - z}\, dz \right) d\tau(\lambda) = - \int_a^b \varphi(\lambda)\, d\tau(\lambda).$$

On the other hand,

$$\int_{\Gamma_\varepsilon} \varphi(z) F_1(z)\, dz = \int_{\Delta_\varepsilon} \varphi(z) F_1(z)\, dz + \left(\int_{K_{1\varepsilon}} + \int_{K_{2\varepsilon}} \right) \varphi(z) F_1(z)\, dz.$$

So, it remains to verify that

$$\lim_{\varepsilon \to 0} \int_{K_{1\varepsilon}} \varphi(z) F_1(z)\, dz = \lim_{\varepsilon \to 0} \int_{K_{2\varepsilon}} \varphi(z) F_1(z)\, dz = 0.$$

As the length of $K_{i\varepsilon}$ $(i = 1, 2)$ equals $\varepsilon \pi$, it is sufficient to show that

$$|F_1(z)| = \frac{1}{\varepsilon} o(\varepsilon), \tag{2.12}$$

when $z \in K_{1\varepsilon}$ and $z \in K_{2\varepsilon}$.

Consider, for example, the case where $z \in K_{1\varepsilon}$, that is $z = a + \varepsilon e^{i\varphi}, \frac{\pi}{2} \leq \varphi \leq \frac{3\pi}{2}$. In this situation

$$|F_1(z)| = |F_1(a + \varepsilon e^{i\varphi})|$$

$$\leq \int_a^b \frac{|d\tau(\lambda)|}{\sqrt{\varepsilon^2 + (\lambda - a)^2}} = \left(\int_a^{a+(n-1)\varepsilon} + \int_{a+(n-1)\varepsilon}^b \right) \frac{|d\tau(\lambda)|}{\sqrt{\varepsilon^2 + (\lambda - a)^2}}$$

$$\leq \frac{1}{\varepsilon} \int_a^{a+(n-1)\varepsilon} |d\tau(\lambda)| + \frac{1}{(n-1)\varepsilon} \int_{a+(n-1)\varepsilon}^b |d\tau(\lambda)|$$

$$\leq \frac{1}{\varepsilon} \left(\int_a^{a+(n-1)\varepsilon} |d\tau(\lambda)| + \frac{1}{(n-1)} \int_a^b |d\tau(\lambda)| \right).$$

Since the expression in the brackets becomes arbitrarily small for sufficiently large n and $0 < \varepsilon < \frac{1}{n^2}$, the relation (2.12) is true. Hence, (2.10) is also valid. The equality (2.11) can be proved in the same way if we use the complements of $K_{1\varepsilon}$ and $K_{2\varepsilon}$ to the whole circles.

In the general situation we can always find $\varepsilon > 0$ such that $a + \varepsilon$ and $b + \varepsilon$ are points of continuity for $\tau(\lambda)$, hence the formula (2.9) is valid where $a + \varepsilon$ and $b + \varepsilon$ stand for a and b respectively. It remains to pass to the limit along the suitable sequence $\varepsilon_n \to 0$.   □

2.3   The function $f_M(z)$ considered separately in the upper and lower half-planes gives, generally speaking, two different meromorphic functions which can cease to be analytic continuations of each other.

We shall call the point $a \in \mathbb{R}^1$ a regular point of the function $f_M(z)$ if the two functions corresponding to $f_M(z)$ admit analytic continuation into each other through some interval $(a - \delta, a + \delta)$, $\delta > 0$. The function $f_M(z)$ is regular on a set if every point of this set is regular for $f_M(z)$.

Theorem 2.5 *Let A $(\overline{\mathcal{D}(A)} = \mathfrak{H})$ be a simple Hermitian operator in $\mathfrak{H}$, whose deficiency index is $(1,1)$. Suppose that the element $f \in \mathfrak{H}$ is such that the function $f_M(z)$ corresponding to f under the map (2.6) is regular on the closed interval $\Delta = [a, b]$. Then for any spectral function E_λ of the operator A the equality*

$$E_\Delta f = \int_a^b f_u(\lambda) \, dE_\lambda u \tag{2.13}$$

holds.

Proof. As has been stated in subsection 1.4

$$E_\lambda = P\widetilde{E}_\lambda, \quad \lambda \in \mathbb{R}^1, \tag{2.14}$$

where $\widetilde{E}_\lambda$ is the spectral function of a certain self-adjoint extension $\widetilde{A}$ to the space $\widetilde{\mathfrak{H}} \supseteq \mathfrak{H}$ and P the orthoprojector from $\widetilde{\mathfrak{H}}$ onto $\mathfrak{H}$. Having arbitrarily fixed a non-real number ζ and a vector $\psi_0 \neq 0$, $\psi_0 \in \mathfrak{N}_\zeta \subset \widetilde{\mathfrak{N}}_\zeta$, we construct the vector-function

$$\psi(z) = \widetilde{U}_{\zeta z}\psi_0 = \psi_0 + (z - \zeta)\,\widetilde{R}_z\psi_0 = \int\limits_{-\infty}^{\infty} \frac{\lambda - \zeta}{\lambda - z}\,d\widetilde{E}_\lambda\psi_0, \quad \Im z \neq 0. \tag{2.15}$$

In view of property 4° of the operator $\widetilde{U}_{\zeta z}$, $\psi(\overline{z})$ is orthogonal to $\mathfrak{M}_z$ for any $z : \Im z \neq 0$. It follows that

$$\left(f - f_M(z), \psi(\overline{z})\right) = 0, \quad f \in \mathfrak{H}.$$

Thus, for each $f \in \mathfrak{H}$,

$$\left(f, \psi(\overline{z})\right) = f_u(z)\left(u, \psi(\overline{z})\right), \quad \Im z \neq 0. \tag{2.16}$$

Because of (2.14) and (2.15),

$$(f, \psi(\overline{z})) = \int\limits_{-\infty}^{\infty} \frac{\lambda - \overline{\zeta}}{\lambda - z}\,d\left(\widetilde{E}_\lambda f, \psi_0\right)$$

$$= \int\limits_{-\infty}^{\infty} \frac{\lambda - \overline{\zeta}}{\lambda - z}\,d\left(E_\lambda f, \psi_0\right).$$

Assume first that the numbers a and b are points of continuity for the function E_λ. Applying Theorem 2.4 to the cases of

$$d\tau(\lambda) = (\lambda - \overline{\zeta})\,d\left(E_\lambda f, \psi_0\right), \quad \varphi(\lambda) = (\lambda - \overline{\zeta})^{-1}$$

and

$$d\tau(\lambda) = (\lambda - \overline{\zeta})\,d\left(E_\lambda u, \psi_0\right), \quad \varphi(\lambda) = f_u(\lambda)(\lambda - \overline{\zeta})^{-1},$$

we find that

$$\lim_{\varepsilon \to 0} \int\limits_{\Delta_\varepsilon} \frac{(f, \psi(\overline{z}))}{z - \overline{\zeta}}\,dz = -\int\limits_a^b d\left(E_\lambda f, \psi_0\right) = -\left(E_\Delta f, \psi_0\right)$$

and

$$\lim_{\varepsilon \to 0} \int\limits_{\Delta_\varepsilon} \frac{f_u(z)}{z - \overline{\zeta}}\,(u, \psi(\overline{z}))\,dz = -\int\limits_a^b f_u(\lambda)\,d\left(E_\lambda u, \psi_0\right)$$

respectively. According to (2.16)

$$\left(E_\Delta f, \psi_0\right) \;=\; \int_a^b f_u(\lambda)\, d\left(E_\lambda u, \psi_0\right); \tag{2.17}$$

here the element ψ_0 from $\mathfrak{N}_\zeta$ and the number $\zeta : \Im\zeta \neq 0$ are arbitrary. Since the closed linear span (c.l.s.) of all $\mathfrak{N}_\zeta$, $\Im\zeta \neq 0$ coincides with $\mathfrak{H}$ (the operator A is simple), the equality (2.17) implies (2.13).

   We have considered the situation when a and b are the points of continuity for E_λ. In the general case there exists $\varepsilon > 0$ as small as desired and such that $a+\varepsilon$ and $b+\varepsilon$ are points of continuity for E_λ. Hence, (2.13) holds where a and b are replaced by $a+\varepsilon$ and $b+\varepsilon$ respectively. Passing to the limit along the suitable $\varepsilon_n \to 0$, we obtain (2.13) in the general situation. $\square$

Corollary 2.1 *Let E_λ be a spectral function of a simple Hermitian operator A with dense domain, and*

$$\sigma(\lambda) \;=\; \left(E_\lambda u, u\right).$$

If the functions $g_M(z)$ and $f_M(z)$ are regular on the closed interval $\Delta = [a, b]$ for some elements $g, f \in \mathfrak{H}$, then

$$\left(E_\Delta f, g\right) \;=\; \int_a^b f_u(\lambda)\overline{g_u(\lambda)}\, d\sigma(\lambda). \tag{2.18}$$

Proof. By (2.13),

$$\left(E_\Delta f, g\right) \;=\; \int_a^b f_u(\lambda)\, d\left(E_\lambda u, g\right) \;=\; \int_a^b f_u(\lambda)\, d\left(u, E_\lambda g\right). \tag{2.19}$$

On the other hand, taking in (2.13) $\Delta = [a, \lambda]$ and $f = g$, we get

$$\left(u, E_\lambda g\right) = \left(u, E_a\right) + \int_a^\lambda \overline{g_u(\lambda)}\, d\left(u, E_\lambda u\right) = \int_a^\lambda \overline{g_u(\lambda)}\, d\sigma(\lambda). \tag{2.20}$$

   The comparison of (2.19) and (2.20) yields (2.18). $\square$

3 On the Spectrum of Self-adjoint Extensions of a Hermitian Operator whose Deficiency Index is (1,1)

3.1 A point z of the complex plane $\mathbb{C}^1$ is called a point of regular type of a linear operator A if there exists a number $\kappa_z > 0$ such that the inequality

$$\|Af - zf\| \geq \kappa_z \|f\| \tag{3.1}$$

holds for all $f \in \mathcal{D}(A)$.

The set of all points of regular type of the operator A is always open. It is not difficult to see that if a linear operator A is closed and z is its point of regular type, then the set $\mathfrak{M}_z = (A - zI)\mathcal{D}(A)$ is closed. Suppose that a point λ is not an eigenvalue of A. Then the closedness of $\mathfrak{M}_\lambda$ is not only a necessary but also a sufficient condition for λ to be a point of regular type for this operator.

In fact, if A is closed, then $\mathcal{D}(A)$ forms a complete normed space with respect to the graph norm of the operator $A - \lambda I$. This operator maps the Banach space $\mathcal{D}(A)$ bijectively and continuously onto the complete Hilbert space $\mathfrak{M}_\lambda$. By the well-known Banach theorem the inverse of the operator $A - \lambda I$ is also continuous. So, there exists a constant $\kappa > 0$ such that

$$\|(A - \lambda I)f\| \geq \kappa \|f\|_{A - \lambda I}.$$

Since $\|f\|_{A - \lambda I} \geq \|f\|$, we have

$$\|(A - \lambda I)f\| \geq \kappa \|f\|,$$

that is, λ is a point of regular type for the operator A, which had to be proved.

If A is a closed Hermitian operator on $\mathfrak{H}$, then the dimension of the subspace $\mathfrak{N}_\lambda = \mathfrak{H} \ominus \mathfrak{M}_{\bar{\lambda}}$ is fixed when λ runs through an arbitrary 1-connected component in the set of points of regular type of the operator A (see, for instance, Ahiezer and Glazman [1]). So, if a closed Hermitian operator has at least one real point of regular type, then its defect numbers are equal.

Definition 3.1 *A finite interval $(a, b) \subset \mathbb{R}^1$ is called a spectral lacuna of a Hermitian operator A if*

$$\left\|\left(A - \frac{b - a}{2} I\right)f\right\| \geq \frac{b - a}{2} \|f\|, \quad f \in \mathcal{D}(A).$$

Consequently, the inequality (3.1) for a real point $z = a$ means that the interval $\mathcal{I}_a = (a - \kappa_a, a + \kappa_a)$ is a spectral lacuna for the operator A.

Proposition 3.1 *Let the interval $\mathcal{I}_a = (a - \kappa_a, a + \kappa_a)$ be a spectral lacuna for a closed Hermitian operator A. Then there exists at least one self-adjoint extension $\widetilde{A}$ within $\mathfrak{H}$ of the operator A such that $\widetilde{A}$ has no points of spectrum in $\mathcal{I}_a$.*

Indeed, taking the operator $\frac{1}{\kappa_a}(A - aI)$ for A, we will arrive at the case of $a = 0$, $\kappa_a = 1$. We may at once assume without loss of generality that $\mathcal{I}_a = \mathcal{I}_0 = (-1, 1)$. Thus, the operator A satisfies the condition

$$\|Af\| \geq \|f\|, \quad f \in \mathcal{D}(A). \tag{3.2}$$

The inequality (3.2) shows that the Hermitian operator A^{-1} is bounded on the subspace $\mathcal{R}(A)$ ($\mathcal{R}(\cdot)$ denotes the range of an operator) and $\|A^{-1}\| \leq 1$. Hence (see Ahiezer and Glazman [1]), A^{-1} admits at least one self-adjoint extension $\widetilde{A^{-1}}$ to the whole space $\mathfrak{H}$. The operator $\widetilde{A} = \left(\widetilde{A^{-1}}\right)^{-1}$ is a self-adjoint extension of A. As $\|\widetilde{A^{-1}}g\| \leq \|g\|$, the inequality $\|\widetilde{A}f\| \geq \|f\|$ is fulfilled for $f \in \mathcal{D}(\widetilde{A})$, which completes the proof.

Proposition 3.2 *Let A be a closed Hermitian operator on $\mathfrak{H}$, whose deficiency index is $(1,1)$, and $\mathcal{I}_a$ its spectral lacuna. Then any of its self-adjoint extensions within $\mathfrak{H}$ has no points of spectrum in the interval $\mathcal{I}_a$ other than one eigenvalue of multiplicity 1.*

To prove this we may be restricted again by the case of $\mathcal{I}_a = (-1, 1)$ and $\|Af\| \geq \|f\|$, $f \in \mathcal{D}(A)$. As has been mentioned in subsection 1.2, the continuous spectra of all self-adjoint extensions within $\mathfrak{H}$ of the operator A coincide. By Proposition 3.1 only eigenvalues of multiplicity not more than 1 can appear in (-1,1) for any such an extension.

Suppose that a self-adjoint extension $\widetilde{A}$ has two eigenvalues λ_1 and λ_2. Then

$$\dim_{\mathcal{D}(A)} \mathcal{D}(\widetilde{A}) > 1$$

because the orthonormal eigenvectors φ_1 and φ_2 of the operator $\widetilde{A}$, corresponding to λ_1 and λ_2 respectively, are linearly independent modulo $\mathcal{D}(A)$, that is the relation $\alpha_1\varphi_1 + \alpha_2\varphi_2 \in \mathcal{D}(A)$ implies $\alpha_1 = \alpha_2 = 0$. Indeed, under the conditions on the operator A,

$$\|A(\alpha_1\varphi_1 + \alpha_2\varphi_2)\|^2 \geq \|\alpha_1\varphi_1 + \alpha_2\varphi_2\|^2 = |\alpha_1|^2 + |\alpha_1|^2.$$

On the other hand, in view of $|\lambda_1| < 1$ and $|\lambda_2| < 1$,

$$\|A(\alpha_1\varphi_1 + \alpha_2\varphi_2)\|^2 = |\lambda_1|^2|\alpha_1|^2 + |\lambda_2|^2|\alpha_2|^2 < |\alpha_1|^2 + |\alpha_2|^2,$$

which is possible only if $\alpha_1 = \alpha_2 = 0$.

Proposition 3.3 *In order that $a \in \mathbb{R}^1$ be a point of regular type for a closed simple Hermitian operator A whose deficiency index is $(1,1)$, it is necessary and sufficient that there exist at least one self-adjoint extension $\widetilde{A}$ within $\mathfrak{H}$ of the operator A such that $\widetilde{A}$ has no points of spectrum other than a finite number of eigenvalues of finite multiplicity in some neighborhood $\mathcal{I}_a$ of the point a.*

The necessity has been established above. Assume that a self-adjoint extension $\widetilde{A}$ satisfies the conditions of the proposition. If no eigenvalue of $\widetilde{A}$ from $\mathcal{I}_a$ coincides with a, then a is a regular point of the operator $\widetilde{A}$, and hence, a point of regular type for A. But if the point a coincides with one of these eigenvalues, then a is not an eigenvalue of any other self-adjoint extension within $\mathfrak{H}$ of the operator A and therefore a is a point of regular type for A. Indeed, suppose that the point a, being an eigenvalue of $\widetilde{A}$, is also an eigenvalue of another self-adjoint extension $\widetilde{A}_1$ within $\mathfrak{H}$ of A. Let φ and φ_1 be the corresponding orthonormal eigenvectors:

$$\widetilde{A}\varphi = a\varphi, \quad \widetilde{A}_1\varphi_1 = a\varphi_1.$$

Since the operator A is simple, the vectors φ and φ_1 do not belong to $\mathcal{D}(A)$. By virtue of (1.8), (1.9),

$$\varphi = \varphi_0 + \varphi_{\bar{z}} + U\varphi_{\bar{z}} \tag{3.3}$$

and

$$\varphi_1 = \varphi_0^1 + c\left(\varphi_{\bar{z}} + U_1\varphi_{\bar{z}}\right), \quad \Im z \neq 0, \quad c = \text{const}, \tag{3.4}$$

where $\varphi_0, \varphi_0^1 \in \mathcal{D}(A), \varphi_{\bar{z}} \in \mathfrak{N}_{\bar{z}}, U$ and U_1 are unitary operators from $\mathfrak{N}_{\bar{z}}$ onto $\mathfrak{N}_z$. Taking into account that the vectors φ and φ_1 are solutions of the equation $A^*u = au$ and that $\dim \mathfrak{N}_a = 1$, we can always consider $\varphi = \varphi_1$. Then the equality

$$\varphi_{\bar{z}} + U\varphi_{\bar{z}} = c\left(\varphi_{\bar{z}} + U_1\varphi_{\bar{z}}\right)$$

follows from (3.3) and (3.4), whence $c = 1$ and $U_1 = U$. Thus, $\widetilde{A} = \widetilde{A}_1$, which completes the proof.

Proposition 3.4 *Let A be a closed simple Hermitian operator in $\mathfrak{H}$ whose deficiency index is (1,1). Suppose that the interval $\mathcal{I}$ of the real axis consists of points of regular type of the operator A. Then the eigenvalues of two different self-adjoint extensions $\widetilde{A}_1$ and $\widetilde{A}_2$ within $\mathfrak{H}$ of the operator A alternate.*

In fact, let the points λ, μ be two neighbouring eigenvalues of the operator $\widetilde{A}_1$ in the interval $\mathcal{I}$. Assume that there are no eigenvalues of the operator $\widetilde{A}_2$ in the interval (λ, μ). Reasoning as in the proof of Propositions 3.2 and 3.3, one can prove that λ and μ are regular points of the operator $\widetilde{A}_2$. Therefore, for a sufficiently small $\varepsilon > 0$,

$$\left\|\left(\widetilde{A}_2 - \frac{\lambda+\mu}{2}\,I\right)f\right\|^2 = \int_{-\infty}^{\lambda-\varepsilon}\left(t - \frac{\lambda+\mu}{2}\right)^2 d\left(E_t f, f\right) + \int_{\mu+\varepsilon}^{\infty}\left(t - \frac{\lambda+\mu}{2}\right)^2 d\left(E_t f, f\right)$$

$$\geq \frac{(\mu - \lambda + 2\varepsilon)^2}{4}\left(\int_{-\infty}^{\lambda-\varepsilon} d\left(E_t f, f\right) + \int_{\mu+\varepsilon}^{\infty} d\left(E_t f, f\right)\right) = \frac{(\mu - \lambda + 2\varepsilon)^2}{4}\,\|f\|^2,$$

where $E_t, t \in \mathbb{R}^1$, is the spectral function of the operator $\tilde{A}_2, f \in \mathcal{D}(\tilde{A}_2)$. So, the interval $(\lambda - \varepsilon, \mu + \varepsilon)$ is a spectral lacuna of the operator $\tilde{A}_2$, hence of the operator A, contrary to Proposition 3.2. Thus, there is at least one eigenvalue of the operator $\tilde{A}_2$ in the interval (λ, μ). If the operator A had two eigenvalues $p, q \in (\lambda, \mu)$, it would be possible to conclude that there is at least one eigenvalue of the operator $\tilde{A}_1$ between p and q, contrary to the fact that λ and μ are neighbouring eigenvalues of this operator.

Proposition 3.5 *If $a \in \mathbb{R}^1$ is a point of regular type for a closed Hermitian operator A, whose deficiency index is (m, m), $m < \infty$, then there exists a self-adjoint extension $\tilde{A}$ within $\mathfrak{H}$ of A such that the point a is its eigenvalue.*

Indeed, let $\mathfrak{N}_a$ be the eigensubspace corresponding to the eigenvalue a of the operator A^*. As was mentioned above, its dimension is equal to m. Define the operator $\tilde{A}$ on $\mathfrak{H}$ as

$$\tilde{A} f = A^* f, \quad \mathcal{D}(\tilde{A}) = \mathcal{D}(A) \dotplus \mathfrak{N}_a.$$

It is not difficult to verify that the operator $\tilde{A}$ is Hermitian. Since $\dim_{\mathcal{D}(A)} \mathcal{D}(\tilde{A}) = m$, the operator $\tilde{A}$ is self-adjoint. Obviously, $\mathfrak{N}_a$ is an eigensubspace of the extension $\tilde{A}$.

The spectrum of a self-adjoint operator B is said to be discrete if in every finite interval it consists of a finite number of eigenvalues of finite multiplicity.

The following assertion results from Propositions 3.2–3.5.

Proposition 3.6 *Let A be a closed simple Hermitian operator whose deficiency index is (m, m), $m < \infty$. If the spectrum of at least one self-adjoint extension within $\mathfrak{H}$ of the operator A is discrete, then the spectrum of any such an extension is also discrete.*

For this to be the case it is necessary and sufficient that every point of the real axis be a point of regular type for the operator A. If this is the case and $m = 1$, then a certain self-adjoint extension $\tilde{A}_\xi$ within $\mathfrak{H}$ of the operator A corresponds to a number $\xi \in \mathbb{R}^1$ so that the point ξ is an eigenvalue of the operator $\tilde{A}_\xi$. Moreover, if ξ passes through the whole real axis, then $\tilde{A}_\xi$ runs through the set of all self-adjoint extensions within $\mathfrak{H}$ of the operator A.

Definition 3.2 *A closed Hermitian operator is called regular if any point of the complex plane is its point of regular type.*

For the kind of simple operator whose deficiency index is $(m, m), m < \infty$, the regularity condition is equivalent to discreteness of the spectrum of every self-adjoint extension within $\mathfrak{H}$ of this operator.

4 Some Aspects of the Theory of Analytic Functions

In this section some results of the theory of analytic functions are given which will be used when constructing a representation of entire operators.

4.1 We start from one generalization of a harmonic function.

Definition 4.1 *A real function $h(x, y)$ is called subharmonic in a domain $G \subset \mathbb{R}^2$ if it satisfies the following conditions:*

1) *the function $h(x, y)$ is given and continuous at all points of G except, possibly, a finite number of points or a sequence $\{(x_n, y_n)\}_{n=1}^{\infty}$ which has no limit points inside G; moreover, at every exceptional point (x_n, y_n),*

$$\lim_{(x,y) \to (x_n, y_n)} h(x, y) = -\infty;$$

it therefore makes sense to put

$$h(x_n, y_n) = -\infty;$$

2) *for each point $(x, y) \in G$ and a sufficiently small ρ,*

$$h(x, y) \leq \frac{1}{2\pi} \int_0^{2\pi} h(x + \rho \cos \alpha, y + \rho \sin \alpha) \, d\alpha. \tag{4.1}$$

The following elementary properties result from the definition of a subharmonic function:

(i) The multiplication of a subharmonic function in the domain G by a positive constant gives a function subharmonic in G.

(ii) The sum of two subharmonic functions in G is a function subharmonic in G.

(iii) The upper envelope

$$u(x, y) \; = \; \sup_t \left\{ u_t(x, y) \right\}$$

of a family (finite or infinite) of subharmonic functions in G is a function subharmonic in G, if the function $u(x, y)$ is continuous everywhere in G except, possibly, a countable set of the points where $u(x, y)$ is equal to $-\infty$.

Let us clarify this property. Indeed, in the opposite case it is possible to find a point $(x_0, y_0) \in G$ such that

$$u(x_0, y_0) > \frac{1}{2\pi} \int_0^{2\pi} u(x + \rho \cos \alpha, y + \rho \sin \alpha) \, d\alpha.$$

As $u(x, y) = \sup_t u_t(x, y)$, there exists a function $u_{t_0}(x, y)$ such that the inequality

$$u_{t_0}(x_0, y_0) > \frac{1}{2\pi} \int_0^{2\pi} u(x + \rho\cos\alpha, y + \rho\sin\alpha)\, d\alpha$$

holds. Because $u(x, y) \geq u_{t_0}(x, y)$, we have

$$u_{t_0}(x_0, y_0) > \frac{1}{2\pi} \int_0^{2\pi} u_{t_0}(x + \rho\cos\alpha, y + \rho\sin\alpha)\, d\alpha,$$

contrary to the fact that the function $u_{t_0}(x, y)$ is subharmonic in G.

(iv) The limit of a uniformly converging sequence of functions subharmonic in G is a subharmonic function in G.

In fact, the limit of a uniformly converging sequence of continuous functions is a continuous function. If, in addition, the inequality (4.1) is fulfilled for any function from the given sequence, then it is also valid for the limit function due to the possibility of the passage to the limit under the integral sign.

A simple example of a subharmonic function is a harmonic one. Another example is $|f(z)|$, where $f(z)$ is an analytic function.

Let a function $f(z)$ be analytic in a domain G and $a \in G$. Surround the point a by a disk $|z - a| < r$ lying inside G, so that the function $f(z)$ has no zeroes on the boundary of this disk. Then the Jensen formula

$$\frac{1}{2\pi} \int_0^{2\pi} \log|f(a + re^{i\varphi})|\, d\varphi = \log\left|\frac{f^{(\lambda)}(a)}{\lambda!}\right| + \log\left(r^\lambda \prod_{k=1}^{n(r)} \frac{r}{|\alpha_k - a|}\right) \qquad (4.2)$$

holds (see, for instance, A.I. Markushevich [1]), where α_k are zeroes different from a of the function $f(z)$ in the disk $|z - a| < r$, $n(r)$ is the number of these zeroes, λ is the multiplicity of the zero a.

Note that this formula can be written in another form. Namely, it is simple to verify that $(a = 0)$

$$\sum_{k=1}^{n(R)} \log\frac{R}{|a_k|} = \int_0^R \frac{n(r)}{r}\, dr.$$

Denote the last integral by $N(R)$. Then

$$N(r) = \frac{1}{2\pi} \int_0^{2\pi} \log|f(re^{i\varphi})|\, d\varphi - \lambda\log r - \log\frac{|f^{(\lambda)}(0)|}{\lambda!}. \qquad (4.3)$$

It follows from the formula (4.2) that

$$\log |f(a)| \le \frac{1}{2\pi} \int_0^{2\pi} \log \left| f\left(a + re^{i\varphi}\right) \right| d\varphi, \tag{4.4}$$

when $f(a) \ne 0$. Since $\log |f(a)| = -\infty$ when $f(a) = 0$, the inequality (4.4) is valid in this situation too. Hence, the function $\log |f(z)|$ is subharmonic in G.

Put

$$\log^+ \alpha = \max \{\log \alpha, 0\} = \begin{cases} \log \alpha & \text{if} \quad \alpha \ge 1 \\ 0 & \text{if} \quad 0 < \alpha < 1 \end{cases}.$$

and

$$\log^- \alpha = \max \{-\log \alpha, 0\} = \log^+ \frac{1}{\alpha}.$$

Then

$$\log \alpha = \log^+ \alpha - \log^- \alpha.$$

Since the maximum of two subharmonic functions is, by property (iii), a subharmonic function, the function $\log^+ |f(z)| = \max \{\log |f(z)|, 0\}$ is subharmonic in the domain G if $f(z)$ is analytic in G.

The following generalized maximum principle holds for subharmonic functions (Markushevich [1]). Let $h(x, y) = h(z)$, $z = x + iy$, be a subharmonic function in a domain G and $u(z)$ a harmonic one. If

$$\varlimsup_{z \to \zeta} \left(h(z) - u(z) \right) \le 0$$

for every boundary point of the domain G except, possibly, a finite number of points $\zeta_1, \ldots, \zeta_m$, and

$$\varlimsup_{z \to \zeta} \left(h(z) - u(z) \right) < +\infty, \quad \text{then} \quad h(z) \le u(z)$$

everywhere inside G. The equality sign at some interior point of G in the last relation holds if and only if $h(z) \equiv u(z)$.

In short, but not quite precisely, this principle may be formulated as follows: if a harmonic function $u(z)$ exceeds a subharmonic one $h(z)$ at all boundary points of a domain G except, possibly, a finite number of them, then $u(z)$ exceeds $h(z)$ inside G.

The analogue of the Phragmen-Lindelöf theorem is also valid for subharmonic functions (Privalov [1]). It says that if a function $u(z)$ subharmonic inside the angle $G = \{z = re^{i\varphi} : |\varphi| < \alpha\frac{\pi}{2}, 0 < r < +\infty\}$ satisfies the following conditions,

1) for every finite point ζ lying on the sides of the angle G

$$u(\zeta) \le c = \text{const};$$

2) whatever $\varepsilon > 0$, the inequality

$$u(z) < \varepsilon r^{1/\alpha}$$

is fulfilled for sufficiently large r, then

$$u(z) \le c$$

at any point from G. Moreover, the equality sign at some point $z_0 \in G$ in this relation is possible if and only if $u(z) \equiv c$.

4.2   The following class of analytic functions plays an important role in the spectral theory of operators.

Definition 4.2 *We shall call a function $f(z)$ an N-function in a domain G if it is analytic inside G and the function $\log^+ |f(z)|$ has a harmonic majorant there. The last condition means that there exists at least one function $u(z)$ harmonic in G such that*

$$\log^+ |f(z)| \le u(z), \quad z \in G.$$

It is evident that under a conformal mapping every N-function in G is transformed into a certain N-function in the image G_1 of the domain G under this mapping.

In the case where a function $f(z)$ is analytic in the disk $|z| < R$, the fact that $f(z)$ is an N-function in this disk is equivalent to the inequality

$$\sup_{0 \le r < R} \int_0^{2\pi} \log^+ \left| f\left(re^{i\varphi}\right) \right| d\varphi < \infty. \tag{4.5}$$

Indeed, suppose that $f(z)$ is an N-function in $G = \{z : |z| < R\}$. Since $\log^+ |f(z)|$ has a harmonic majorant $u(z)$, we have

$$\int_0^{2\pi} \log^+ \left| f\left(re^{i\varphi}\right) \right| d\varphi \le \int_0^{2\pi} u\left(re^{i\varphi}\right) d\varphi = 2\pi u(0).$$

Conversely, let the inequality (4.5) hold. Choose an increasing number sequence $r_n \to R$ as $n \to \infty$ and construct the sequence of functions $u_n(z)$ harmonic in the disks $|z| \le r_n$ respectively so that

$$u_n(z) = \log^+ |f(z)| \quad \text{if } |z| = r_n.$$

By virtue of the generalized maximum principle for subharmonic functions (see subsection 4.1), the function $u_n(z)$ is a harmonic majorant for $\log^+ |f(z)|$ in the disk $|z| \le r_n$. The sequence $u_n(z)$ increases monotonically. Since $u_{n+1}(z) \ge \log^+ |f(z)|$ for $|z| \le r_{n+1}$ and $\log^+ |f(z)| = u_n(z)$ as $|z| = r_n$, the inequality $u_{n+1}(z) \ge u_n(z)$ holds for the points $z : |z| = r_n$. But the minimum of a harmonic function may be attained only on the boundary of a domain. So, $u_{n+1}(z) - u_n(z) \ge 0$ inside the disk $|z| < r_n$. In addition, the number sequence

$$u_n(0) = \frac{1}{2\pi} \int_0^{2\pi} u_n(re^{i\varphi}) \, d\varphi = \frac{1}{2\pi} \int_0^{2\pi} \log^+ |f(re^{i\varphi})| \, d\varphi$$

is bounded. By the Harnack theorem (see, for instance, Markushevich [1]), the sequence $u_n(z)$ converges uniformly inside the disk $|z| < R$ to a certain harmonic function $u(z)$ which is a majorant for $\log^+ |f(z)|$.

It should be noted that for a function $f(z)$ analytic in the disk $|z| < R$ and different from the identical zero, the condition (4.5) is equivalent to the inequality

$$\varlimsup_{r \to R} \int_0^{2\pi} \left| \log |f(re^{i\varphi})| \right| \, d\varphi < \infty. \tag{4.6}$$

Indeed, by the Jensen formula (4.2),

$$\frac{1}{2\pi} \int_0^{2\pi} \log^+ |f(re^{i\varphi})| \, d\varphi - \frac{1}{2\pi} \int_0^{2\pi} \log^- |f(re^{i\varphi})| \, d\varphi$$

$$= \log \left| \frac{f^{(\lambda)}(0)}{\lambda!} \right| + \lambda \log r + \sum_{k=1}^{n(r)} \log \frac{r}{|\alpha_k|},$$

whence

$$\frac{1}{2\pi} \int_0^{2\pi} \log^- |f(re^{i\varphi})| \, d\varphi$$

$$= \frac{1}{2\pi} \int_0^{2\pi} \log^+ |f(re^{i\varphi})| \, d\varphi - \log \frac{|f^{(\lambda)}(0)|}{\lambda!} - \lambda \log r - \sum_{k=1}^{n(r)} \log \frac{r}{|\alpha_k|}.$$

Therefore, the boundedness of the integral

$$\frac{1}{2\pi} \int_0^{2\pi} \log^+ |f(re^{i\varphi})| \, d\varphi$$

implies that of the integral

$$\frac{1}{2\pi} \int_0^{2\pi} \log^- \left| f\left(re^{i\varphi}\right) \right| d\varphi$$

for r sufficiently close to R. Thus,

$$\varlimsup_{r \to R} \int_0^{2\pi} \left| \log \left| f\left(re^{i\varphi}\right) \right| \right| d\varphi$$

$$= \varlimsup_{r \to R} \left(\int_0^{2\pi} \log^+ \left| f\left(re^{i\varphi}\right) \right| d\varphi + \int_0^{2\pi} \log^- \left| f\left(re^{i\varphi}\right) \right| d\varphi \right) < \infty.$$

On the other hand, it is obvious that

$$\int_0^{2\pi} \log^+ \left| f\left(re^{i\varphi}\right) \right| d\varphi \leq \int_0^{2\pi} \left| \log \left| f\left(re^{i\varphi}\right) \right| \right| d\varphi.$$

It follows from this estimate that the relation (4.6) implies (4.5).

Using (4.2), we can also conclude that if $f(z) \not\equiv 0$ is an N-function in the disk $|z| < R$, then

$$\sum_{k=0}^{\infty} (R - |\alpha_k|) < \infty \tag{4.7}$$

where α_k, $k = 0, 1, \ldots$, are the zeroes of $f(z)$ in the disk $|z| < R$.

In fact, the formula (4.2) shows that the mean value of the function $\log |f(z)|$ is a non-decreasing function of the radius of a circle along which this mean is taken. Therefore,

$$\lim_{r \to R} \frac{1}{2\pi} \int_0^{2\pi} \log |f\left(re^{i\varphi}\right)| d\varphi$$

always exists. Because of (4.6) this limit is finite, which, due to (4.2), is equivalent to the boundedness above of

$$\log \prod_{k=1}^{n(r)} \frac{r}{|\alpha_k|}, \quad \frac{R}{2} \leq r < R.$$

The latter, in turn, is equivalent to convergence of the product

$$\prod_{k=1}^{n} \frac{R}{|\alpha_k|} = \prod_{k=1}^{n} 1 + \frac{R - |\alpha_k|}{|\alpha_k|}$$

(in the case of infinite n), where $n \leq \infty$ is the number of the zeroes of the function $f(z)$ in the disk $|z| < R$, hence to convergence of the series

$$\sum_{k=1}^{n} \frac{R - |\alpha_k|}{|\alpha_k|}. \tag{4.8}$$

Since $\frac{R}{2} \leq |\alpha_k| < R$ for sufficiently large k, the series (4.8) converges if and only if the sum $\sum_{k=1}^{n} R - |\alpha_k|$ of the distances from the zeroes of $f(z)$ to the circle $|z| = R$ is finite, hence (4.7) is true.

The set of all N-functions in the disk $|z| < R$ forms a linear ring. Moreover, if the functions $f(z)$ and $g(z)$ belong to this ring and the function $p(z) = \dfrac{f(z)}{g(z)}$ is analytic in the disk $|z| < R$, then $p(z)$ belongs to this ring too. The proof follows from (4.5) and the relations

$$\log^+ |f(z) \pm g(z)| \leq \log 2 + \log^+ |f(z)| + \log^+ |g(z)| \tag{4.9}$$

and

$$\log^+ |f(z)g(z)| \leq \log^+ |f(z)| + \log^+ |g(z)|. \tag{4.10}$$

We shall first prove the inequality (4.9). If $|f(z)| \leq 1, |g(z)| \leq 1$, then

$$\log^+ \left(|f(z)| + |g(z)| \right) < \log 2.$$

But if one of the numbers $|f(z)|$ or $|g(z)|$ is larger than 1, then

$$\log^+ \left(|f(z)| + |g(z)| \right) = \log \left(|f(z)| + |g(z)| \right) < \log \left(2 \max \left\{ |f(z)|, |g(z)| \right\} \right)$$

$$= \log 2 + \log \left(\max \left\{ |f(z)|, |g(z)| \right\} \right) \leq \log 2 + \log^+ |f(z)| + \log^+ |g(z)|.$$

To prove (4.10), suppose first that at least one of the numbers $|f(z)|$ or $|g(z)|$ does not exceed 1. Then

$$\log^+ |f(z)g(z)| \leq \log^+ \max \left\{ |f(z)|, |g(z)| \right\} \leq \log^+ |f(z)| + \log^+ |g(z)|.$$

If $|f(z)| \geq 1, |g(z)| \geq 1$, then

$$\log^+ |f(z)g(z)| = \log |f(z)g(z)| = \log |f(z)| + \log |g(z)| = \log^+ |f(z)| + \log^+ |g(z)|.$$

4.3   Let us consider some properties of N-functions given in the unit disk. The Nevanlinna criterion for $f(z)$ to be such a function will be useful later on.

Theorem 4.1 *A function $f(z)$ given in the unit disk is an N-function in this disk if and only if it admits the representation*

$$f(z) = \varepsilon B(z) \exp \int_{-\pi}^{\pi} \frac{e^{it} + z}{e^{it} - z} d\sigma(t), \quad |z| < 1, \tag{4.11}$$

where $|\varepsilon| = 1, \sigma(t)$ *is a real function of bounded variation on the interval* $[-\pi, \pi]$, $B(z)$ *a Blaschke function, that is a function of the form*

$$B(z) = z^\lambda \prod \frac{\alpha_k - z}{1 - \bar{\alpha}_k z} \frac{|\alpha_k|}{\alpha_k}, \qquad |z| < 1, \tag{4.12}$$

where $\lambda \in \{0\} \cup \mathbb{N}, |\alpha_k| < 1$, *and*

$$\sum_{k=1}^{\infty} (1 - |\alpha_k|) < \infty \tag{4.13}$$

if the product is infinite.

As is known, the condition (4.13) is necessary and sufficient for the product (4.12) to converge. The function $B(z)$ is analytic in the disk $|z| < 1$. Since $|B(z)| = 1$ as $|z| = 1$, we conclude that $|B(z)| < 1$ when $|z| < 1$. The totality of the zeroes within the unit disk of the function $B(z)$ coincides with the set $\underbrace{\{0, \dots, 0}_{\lambda}, \alpha_1, \alpha_2, \dots\}$.

We shall also need the following important result.

Theorem 4.2 *Let a function* $f(z)$ *be analytic in the unit disk. In order that* $f(z)$ *be an N-function, it is sufficient that* $\Re f(z)$ *or* $\Im f(z)$ *retain its sign for* $z : |z| < 1$.

Proof. Suppose, for instance, that $\Re f(z) > 0$. Hence, $-\frac{\pi}{2} < \arg f(z) < \frac{\pi}{2}$. The identity

$$f^\delta(z) = |f(z)|^\delta e^{i\delta \arg f(z)} \quad \text{yields} \quad |f(z)|^\delta \le \frac{\Re f^\delta(z)}{\cos \frac{\delta \pi}{2}},$$

where $\delta \in (0, 1)$ is arbitrary. By the mean value theorem for harmonic functions,

$$\frac{1}{2\pi} \int_0^{2\pi} |f(re^{i\varphi})|^\delta d\varphi \le \frac{\Re f^\delta(0)}{\cos \delta \frac{\pi}{2}}.$$

Consequently,

$$\frac{1}{2\pi} \int_0^{2\pi} |f(re^{i\varphi})|^\delta d\varphi < \infty.$$

The inequality

$$\exp\left(\frac{1}{2\pi} \int_0^{2\pi} \log(1 + |f(re^{i\varphi})|) d\varphi\right) \le \left(\frac{1}{2\pi} \int_0^{2\pi} (1 + |f(re^{i\varphi})|)^\delta d\varphi\right)^{1/\delta} \tag{4.14}$$

enables the conclusion to be drawn that $f(z)$ is an N-function in the unit disk.

The inequality (4.14) can be obtained from the well-known estimate (see Hardy, Littlewood, Polya [1]) for the weighted arithmetic and geometric means of non-negative numbers $g_1, g_2, \ldots, g_n$ and weights $p_1, p_2, \ldots, p_n$,

$$\exp \frac{\sum\limits_{i=1}^{n} p_i \log g_i}{\sum\limits_{i=1}^{n} p_i} = \left(g_1^{p_1} \ldots g_n^{p_n}\right)^{1/\sum\limits_{i=1}^{n} p_i} \leq \frac{\sum\limits_{i=1}^{n} p_i g_i}{\sum\limits_{i=1}^{n} p_i}. \tag{4.15}$$

If we put

$$p(\varphi) \equiv 1, \quad g(\varphi) = \left(1 + \left|f\left(re^{i\varphi}\right)\right|\right)^{\delta}, \quad \varphi \in [0, 2\pi],$$

we will get, passing to the limit in (4.15), the inequality (4.14).   □

It is relevant to remark also that

$$\int\limits_{0}^{1} \int\limits_{0}^{2\pi} \left|\log\left|f\left(re^{i\varphi}\right)\right|\right| r \, d\varphi \, dr < \infty \tag{4.16}$$

for an N-function $f(z), |z| < 1$. This follows from (4.6).

After these tentative comments, let us begin the investigation of the behaviour of N-functions given in the unit disk when approaching its boundary.

Lemma 4.1 *Let* $B(z), |z| < 1$ *be a function of the form* (4.12). *Then there exists an increasing sequence of positive numbers* $r_n < 1, r_n \to 1$, *such that*

$$(1 - r_n) \log \left|B\left(r_n e^{i\theta}\right)\right| \to 0 \quad n \to \infty, \tag{4.17}$$

uniformly with respect to $\theta \in [0, 2\pi]$.

Proof. Set

$$m(r) = \min_{\theta \in [0, 2\pi]} \left|B\left(re^{i\theta}\right)\right|, \quad 0 < r < 1.$$

Since

$$\left|\frac{z - \alpha_k}{1 - \overline{\alpha}_k z}\right|^2 = 1 - \frac{\left(1 - |\alpha_k|^2\right)\left(1 - r^2\right)}{|1 - \overline{\alpha}_k z|^2}$$

$$\geq 1 - \frac{\left(1 - |\alpha_k|^2\right)\left(1 - r^2\right)}{(1 - |\alpha_k| r)^2} = \left(\frac{r - |\alpha_k|}{1 - |\alpha_k| r}\right)^2, \quad r = |z| < 1,$$

we have, according to (4.12), the estimate

$$r^{-\lambda} m(r) \geq |B_1(r)|, \quad \text{in which} \quad B_1(z) = \prod_{k=1}^{\infty} \frac{|\alpha_k| - z}{1 - |\alpha_k| z}.$$

Consider the function

$$\varphi(\zeta) = \prod_{k=1}^{\infty} \left(1 - \frac{\zeta}{\zeta_k}\right), \quad \text{where} \quad \zeta_k = \frac{1 + |\alpha_k|}{1 - |\alpha_k|} > 0, \quad k \in \mathbb{N}.$$

Because

$$\sum_{k=1}^{\infty} \frac{1}{\zeta_k} = \sum_{k=1}^{\infty} \frac{1 - |\alpha_k|}{1 + |\alpha_k|} < \sum_{k=1}^{\infty} (1 - |\alpha_k|) < \infty,$$

the function $\varphi(\zeta)$ is entire of zero genus, hence of minimal type (Markushevich [1]), that is,

$$\varlimsup_{\rho \to \infty} \frac{\log|\varphi(\rho e^{i\theta})|}{\rho} = 0$$

for any $\theta \in [0, 2\pi]$. In particular, in view of the inequality $\varphi(-\rho) > 1$, $\rho > 0$,

$$\lim_{\rho \to \infty} \frac{\log \varphi(-\rho)}{\rho} = 0.$$

So, there exist numbers $\rho_1 < \rho_2 < \cdots,\quad \rho_n \to \infty$, such that

$$\lim_{n \to \infty} \frac{\log|\varphi(\rho_n)|}{\rho_n} = \lim_{n \to \infty} \frac{\log|\varphi(-\rho_n)|}{\rho_n} = 0.$$

On the other hand, it is not difficult to check that

$$B_1(z) = \frac{\varphi(\zeta)}{\varphi(-\zeta)} \qquad \text{if} \qquad z = \frac{\zeta - 1}{\zeta + 1}.$$

Setting

$$r_n = \frac{\rho_n - 1}{\rho_n + 1}, \qquad n \in \mathbb{N},$$

we find that

$$\lim_{n \to \infty} (1 - r_n) \log|B_1(r_n)| = \lim_{n \to \infty} \frac{2}{\rho_n + 1} \log\left|\frac{\varphi(\rho_n)}{\varphi(-\rho_n)}\right| = 0.$$

Since $|B(re^{i\theta})| < 1$ when $0 < r < 1, 0 \leq \theta \leq 2\pi$, the estimates

$$\lambda \log r_n + \log|B_1(r_n)| \leq \log m(r_n) \leq \log|B(r_n e^{i\theta})| < 0, \quad \theta \in [0, 2\pi]$$

hold. Then

$$0 = \lim_{n \to \infty} (1 - r_n)\lambda \log r_n + \lim_{n \to \infty} (1 - r_n)\log|B_1(r_n)|$$
$$\leq \lim_{n \to \infty} (1 - r_n)\log|B(r_n e^{i\theta})| \leq 0,$$

whence

$$\lim_{n \to \infty} (1 - r_n)\log|B(r_n e^{i\theta})| = 0, \quad \theta \in [0, 2\pi]. \qquad \square$$

Lemma 4.2 *Let $B(z)$ be a function of the form (4.12) and Γ a nontangential smooth path inside the unit disk, which leads to the point $z=1$. Then*

$$\varlimsup_{z \in \Gamma, z \to 1} |1 - z| \log|B(z)| = 0, \tag{4.18}$$

Proof. Since $|B(z)| \leq 1$,

$$\varlimsup_{z \to 1} |1 - z| \log |B(z)| \leq 0 .$$

On the other hand, if $\{r_n\}_{n \in \mathbb{N}}$ is the sequence for which the equality (4.17) is true, then choosing numbers $z_n \in \Gamma$ so that $|z_n| = r_n$ (for sufficiently large n), we obtain

$$\lim_{n \to \infty} |1 - z_n| \log |B(z_n)| = \lim_{n \to \infty} \frac{|1 - z_n|}{1 - r_n} \lim_{n \to \infty} (1 - r_n) \log |B(z_n)| = 0$$

because

$$\lim_{n \to \infty} \frac{|1 - z_n|}{1 - r_n} = \frac{1}{\cos \theta}$$

where $\theta = \theta(\Gamma)$ is the angle between the positive x-semiaxis and the tangent to Γ at the point $z = 1$.   □

Theorem 4.3 *Let $F(z)$ be an N-function in the unit disk, Γ and the same as in Lemma 4.2. Then*

$$\varlimsup_{z \in \Gamma, z \to 1} |1 - z| \log |F(z)| = d \cos \theta , \tag{4.19}$$

where the constant d does not depend on the choice of the path Γ, $\theta = \theta(\Gamma)$ is the angle between the positive real semiaxis and the tangent to Γ at $z = 1$.

Proof. According to the representation (4.11)

$$F(z) = \varepsilon B(z) F_1(z) \quad \text{with} \quad F_1(z) = \exp \int_{-\pi}^{\pi} \frac{e^{it} + z}{e^{it} - z} \, d\sigma(t) .$$

Taking into account that

$$\int_{-\pi}^{\pi} \frac{e^{it} + z}{e^{it} - z} \, d\sigma(t) = \frac{d}{2} \frac{1 + z}{1 - z} + \int_{-\pi}^{\pi} \frac{e^{it} + z}{e^{it} - z} \, d\sigma_0(t) ,$$

where d is the doubled jump of the function $\sigma(t)$ at 0, $\sigma_0(t)$ is chosen appropriately, we get the equalities

$$|1 - z| \log |F_1(z)| = |1 - z| \, \Re \left(\frac{d}{2} \frac{1 + z}{1 - z} + \int_{-\pi}^{\pi} \frac{e^{it} + z}{e^{it} - z} \, d\sigma_0(t) \right)$$

$$= -\frac{d}{2} |1 - z| + d \cos \varphi + |1 - z| \, \Re \int_{-\pi}^{\pi} \frac{e^{it} + z}{e^{it} - z} \, d\sigma_0(t) ,$$

in which

$$z = 1 - \rho e^{i\varphi}, \quad \varphi \in \left(-\frac{\pi}{2}, \frac{\pi}{2}\right).$$

It should be noted that

$$\varphi \to 0 \quad \text{as} \quad z \to 1, \quad z \in \Gamma.$$

On the other hand, whatever $\delta : 0 < \delta < \pi$, the relation

$$\varlimsup_{z \in \Gamma, z \to 1} |1 - z| \, \Re \int_{-\pi}^{\pi} \frac{e^{it} + z}{e^{it} - z} \, d\sigma_0(t) \leq \varlimsup_{z \in \Gamma, z \to 1} |1 - z| \left| \int_{-\pi}^{\pi} \frac{e^{it} + z}{e^{it} - z} d\sigma_0(t) \right|$$

$$= \varlimsup_{z \in \Gamma, z \to 1} \left| (1 - z) \int_{-\delta}^{\delta} \frac{e^{it} + z}{e^{it} - z} \, d\sigma_0(t) \right|$$

$$\leq \varlimsup_{z \in \Gamma, z \to 1} 2 \frac{|1 - z|}{1 - |z|} \int_{-\delta}^{\delta} d\sigma_0(t) = \frac{2}{\cos \theta} (\sigma_0(\delta) - \sigma_0(-\delta))$$

is valid. As when δ is arbitrarily small so is the last value,

$$\lim_{z \in \Gamma, z \to 1} |1 - z| \log |F_1(z)| = d \cos \theta.$$

This equality together with (4.18) gives (4.19).   □

Remark 4.1 *It is not difficult to observe from the foregoing arguments that the limit equality*

$$\lim_{\rho \to 0} \rho \log |F_1(1 - \rho e^{i\theta})| = d \cos \theta$$

is fulfilled uniformly with respect to $\theta \in [-\gamma, \gamma]$ where γ is an arbitrary fixed number from the interval $\left(0, \frac{\pi}{2}\right)$. Consequently, for the same θ, (4.19) holds uniformly from above, that is, for any $\varepsilon > 0$ and $\gamma \in \left(0, \frac{\pi}{2}\right)$ there exists ρ_ε such that

$$\rho \log |F(1 - \rho e^{i\theta})| = d \cos \theta + \varepsilon, \quad 0 < \rho < \rho_\varepsilon, \; |\theta| < \gamma.$$

4.4   Now let $f(z)$ be an N-function in the upper half-plane. Consider the function

$$F(\zeta) = f(z),$$

where

$$z = i \frac{1 + \zeta}{1 - \zeta}, \quad \zeta = \frac{z - i}{z + i}. \tag{4.20}$$

This function is an N-function in the unit disk. Observe that the map $\zeta = \zeta(z)$ appearing in (4.20) transforms the ray $z = re^{i\varphi}, \quad r \in [0, \infty)$, into the arc Γ of a certain circle whose ends are situated at the points $z = -1$ and $z = 1$. The angle between the tangent to this arc at the point $z = 1$ and the positive real semiaxis is equal to $\theta(\Gamma) = \frac{\pi}{2} - \varphi$. Taking into account also that

$$\lim_{|z| \to \infty} |1 - \zeta||z| = 2,$$

we obtain from Theorem 4.3 the first statement of the following theorem.

Theorem 4.4 *Let $f(z)$ be an N-function in the half-plane $\Im z > 0$. Then there exists a real constant μ such that*

$$\overline{\lim_{r \to \infty}} \frac{\log |f(re^{i\varphi})|}{r} = \mu \sin \varphi \qquad (4.21)$$

uniformly from above in any sector $\delta \leq \varphi \leq \pi - \delta$ $(0 < \delta < \pi)$. If in addition the function $f(z)$ has a finite number of zeroes inside the angle $|\varphi - \varphi_0| < \delta$, $0 < \delta < \varphi_0 < \pi$, then the limit equality

$$\lim_{r \to \infty} \frac{\log |f(re^{i\varphi})|}{r} = \mu \sin \varphi$$

holds uniformly with respect to $\varphi : |\varphi - \varphi_0| < \delta_1$, for every $\delta_1 < \delta$.

Proof. Only the second assertion of the theorem remains to be proved.

In accordance with the factorization

$$F(\zeta) = \varepsilon B(\zeta) F_1(\zeta),$$

where the N-function $F_1(\zeta)$ in the unit disk has no zeroes in this disk, the function $f(z)$ can be represented in the form

$$f(z) = b(z) f_1(z).$$

Here $f_1(z)$, $\Im z > 0$, is an N-function without zeroes in the half-plane $\Im z > 0$,

$$|b(z)| = \prod_{k=1}^{\infty} \left| \frac{z - z_k}{z - \bar{z}_k} \right|,$$

$\{z_k\}_{k \in \mathbb{N}}$ is the sequence of all zeroes of the function $f(z)$ lying in the half-plane $\Im z > 0$. Now the condition (4.13) means that

$$\sum_{k=1}^{\infty} \Im \frac{1}{z_k} < \infty.$$

By Remark 4.1, there exists the limit

$$\lim_{r \to \infty} \frac{\log |f_1(re^{i\varphi})|}{r} = \mu \sin \varphi \qquad (4.22)$$

uniform in every angle $\gamma \leq \varphi \leq \pi - \gamma$. Since the number of zeroes of $f(z)$ is finite inside the angle $|\varphi - \varphi_0| < \delta$, we have

$$|z - z_k| \geq |z_k| \sin(\delta - \delta_1), \quad |\varphi - \varphi_0| \leq \delta_1$$

for sufficiently large $k > N$. Hence,

$$\left| \frac{z - \bar{z}_k}{z - z_k} \right|^2 - 1 = 4 \frac{\Im z \, \Im z_k}{|z - z_k|^2}$$

$$\leq 4 \frac{|z| \, \Im z_k}{|z_k|^2 \sin^2(\delta - \delta_1)} = -4 \frac{|z|}{\sin^2(\delta - \delta_1)} \Im \frac{1}{z_k}, \quad k > N,$$

when z goes through the angle $|\varphi - \varphi_0| \leq \delta_1$, whence

$$\left| \frac{z - \bar{z}_k}{z - z_k} \right|^2 \leq 1 - 4 \frac{|z|}{\sin^2(\delta - \delta_1)} \Im \frac{1}{z_k}, \quad |\varphi - \varphi_0| \leq \delta_1, \ k > N. \tag{4.23}$$

Choosing for arbitrary fixed $\varepsilon > 0$ a number $N > 0$ so that (4.23) holds and

$$-4 \sin^{-2}(\delta - \delta_1) \sum_{k=N+1}^{\infty} \Im \frac{1}{z_k} < \varepsilon,$$

we conclude that

$$\prod_{k=N+1}^{\infty} \left| \frac{z - \bar{z}_k}{z - z_k} \right| \leq \exp\left(\frac{1}{2} \varepsilon |z|\right), \quad |\varphi - \varphi_0| \leq \delta_1.$$

Thus,

$$\frac{1}{2} \exp\left(-\frac{1}{2} \varepsilon |z|\right) \leq |b(z)| \leq 1, \quad |\varphi - \varphi_0| \leq \delta_1, \quad |z| > R_\varepsilon$$

as R_ε is large enough. This shows that

$$\frac{\log |b(re^{i\varphi})|}{r} \to 0 \quad \text{when} \quad r \to \infty$$

uniformly in the angle $|\varphi - \varphi_0| \leq \delta_1$. Taking into account (4.22), we obtain

$$\lim_{r \to \infty} \frac{\log |f(re^{i\varphi})|}{r} = \mu \sin \varphi, \quad |\varphi - \varphi_0| \leq \delta_1,$$

which completes the proof. □

Lemma 4.3 *If $f(z)$ is an N-function in the half-plane $\Im z > 0$, then*

$$\int_0^{\infty} \int_{-\infty}^{\infty} \frac{|\log |f(z)||}{(1 + |z|)^4} \, dx \, dy < \infty. \tag{4.24}$$

Proof. The function $F(\zeta)$ which corresponds to $f(z)$ under the map (4.20) is an N-function in the unit disk $|\zeta| < 1$. Using the inequality (4.16) after the transformation (4.20), we get

$$4 \int\limits_{0}^{\infty} \int\limits_{-\infty}^{\infty} \frac{|\log |f(z)||}{|z + i|^4} \, dx \, dy < \infty,$$

which implies, since $|z + i| \leq 1 + |z|$, the convergence of the integral (4.24). □

Lemma 4.4 *Let $\sigma(\lambda)$, $\lambda \in \mathbb{R}^1$, be a function of bounded variation. Then the integrals*

$$\mathcal{I}(z) = \int\limits_{-\infty}^{\infty} \frac{d\sigma(\lambda)}{\lambda - z}, \quad \Im z \neq 0, \tag{4.25}$$

and

$$f(z) = \int\limits_{-\infty}^{\infty} \frac{\lambda - i}{\lambda - z} \, d\sigma(\lambda), \quad \Im z \neq 0, \tag{4.26}$$

are N-functions in each of the half-planes $\Im z > 0$ and $\Im z < 0$.

For the function (4.25), the proof follows from Theorem 4.2. The representation

$$f(z) = \operatorname{Var} \sigma + (z - i) \int\limits_{-\infty}^{\infty} \frac{d\sigma(\lambda)}{\lambda - z},$$

and the fact that the set of all N-functions forms a linear ring give the proof for (4.26).

Lemma 4.5 *If the function $f(z)$ is of the form (4.26), then*

$$\frac{4}{\pi} \int\limits_{0}^{\infty} \int\limits_{-\infty}^{\infty} \frac{\log^+ |f(z)|}{(1 + |z|)^4} \, dx \, dy \leq \log^+ \operatorname{Var} \sigma + 3, \quad z = x + iy.$$

Proof. The passage to the variables

$$\zeta = \frac{z - i}{z + i}, \quad e^{it} = \frac{\lambda - i}{\lambda + i}$$

in the integral (4.26) yields

$$f(z) = F(\zeta) = (1 - \zeta) \int\limits_{-\pi}^{\pi} \frac{e^{it}}{e^{it} - \zeta} \, d\omega(t),$$

where $\omega(t) = \sigma(\lambda)$. It follows that

$$\left|F(\zeta)\right| \leq (1 + |\zeta|)\frac{1}{1 - |\zeta|} \operatorname{Var}\omega \leq 2\frac{\operatorname{Var}\sigma}{1 - |\zeta|}, \qquad |\zeta| < 1;$$

hence,

$$\log^+ \left|F(\zeta)\right| \leq \log^+ \operatorname{Var}\sigma + \log 2 - \log\left(1 - |\zeta|\right).$$

Therefore,

$$\frac{4}{\pi} \int_0^\infty \int_{-\infty}^\infty \frac{\log^+ |f(z)|}{(1 + |z|)^4}\, dx\, dy$$

$$\leq \frac{4}{\pi} \int_0^\infty \int_{-\infty}^\infty \frac{\log^+ |f(z)|}{|i + z|^4}\, dx\, dy = \frac{1}{\pi} \int_0^1 \int_0^{2\pi} r \log^+ \left|F\left(re^{i\varphi}\right)\right| d\varphi\, dr$$

$$\leq \frac{1}{\pi}\left(\pi \log^+ \operatorname{Var}\sigma + \pi \log 2 - \int_0^1 \int_0^{2\pi} r \log(1 - r)\, d\varphi\, dr\right)$$

$$= \log^+ \operatorname{Var}\sigma + \log 2 + \frac{3}{2} < \log^+ \operatorname{Var}\sigma + 3. \qquad \square$$

The following assertion is of great importance.

Theorem 4.5 *In order that an entire function $f(z)$ be an N-function in each of the half-planes $\Im z > 0$ and $\Im z < 0$, it is necessary and sufficient that it satisfy the conditions*

(i)
$$\int_{-\infty}^\infty \frac{\log^+ |f(x)|}{1 + x^2}\, dx < \infty; \qquad (4.27)$$

(ii) *the function $f(z)$ is of at most exponential type, that is*

$$\varlimsup_{|z|\to\infty} \frac{\log |f(z)|}{|z|} < \infty. \qquad (4.28)$$

Proof. Suppose that an entire function $f(z)$ satisfies the conditions (4.27) and (4.28). Denote by K_R the half-disk lying in the upper half-plane on the interval $[-R, R]$ as on a diameter, and consider the function

$$U_R(z) = \frac{2}{3\pi} \Im \log \frac{R - z}{R + z} = \frac{2}{3\pi} \arg \frac{R - z}{R + z};$$

by arg z we mean here the function

$$
\arg(x+iy) = \begin{cases}
\pi + \arctan \frac{y}{x} & \text{as} \quad x < 0 \quad y \in \mathbb{R}^1 \\
\arctan \frac{y}{x} & \text{as} \quad x > 0 \quad y \geq 0 \\
2\pi + \arctan \frac{y}{x} & \text{as} \quad x > 0 \quad y < 0 \\
\frac{\pi}{2} & \text{as} \quad x = 0 \quad y > 0 \\
\frac{3\pi}{2} & \text{as} \quad x = 0 \quad y < 0.
\end{cases}
$$

The function $U_R(z)$ is harmonic inside K_R and vanishes at $z \in [-R, R]$. It is easy to make sure that $U_R(z)$ possesses the following properties:

a) $U_R(z) \geq 0$ if $z \in K_R$;

b) $U_R(Re^{i\varphi}) = 1$ if $0 < \varphi < \pi$;

c) $U_R(z) = \dfrac{4}{3\pi} \dfrac{y}{R} + O\left(\dfrac{z^2}{R^2}\right), \quad R \to \infty, \ y = \Im z.$

The condition (4.27) makes it possible to introduce the non-negative function

$$
H(z) = \frac{y}{\pi} \int_{-\infty}^{\infty} \frac{\log^+ |f(t)|}{(x-t)^2 + y^2}\, dt, \quad y > 0,
$$

harmonic in the half-plane $\Im z > 0$. Its boundary value $H(x)$ when $y \to 0$, $y > 0$, is continuous, and

$$
H(x) = \log^+ |f(x)| \geq \log|f(x)|, \quad x \in \mathbb{R}^1. \tag{4.29}
$$

By virtue of (4.28), the constants $a > 1$ and $b > 0$ can be found so that

$$
|f(z)| \leq a e^{b|z|}. \tag{4.30}
$$

Then

$$
\log|f(z)| \leq H(z) + bRU_R(z) + \log a, \quad z \in K_R. \tag{4.31}
$$

Indeed, by the generalized maximum principle for subharmonic functions, it suffices to check (4.31) at the boundary of K_R. This boundary consists of the open half-circle Γ_R and the segment $[-R, R]$. On Γ_R the inequality (4.31) is a direct consequence of the property b) of the function $U_R(z)$ and (4.30). On $[-R, R]$ it follows from the estimate (4.29). As R approaches ∞ in (4.31), we arrive at the inequality

$$
\log^+ |f(z)| \leq H(z) + \frac{4}{3\pi} by + \log a, \quad y > 0.
$$

As the function on the right-hand side of this relation is harmonic in the upper half-plane, $f(z)$ is an N-function there. The case of $\Im z < 0$ may be considered in the same way.

Conversely, let an entire function $f(z)$ be an N-function in the half-plane $\Im z > 0$. Then the function $F(\zeta) = f(z)$, $|\zeta| < 1$, where z and ζ are associated with each other by the formulas (4.20), satisfies (4.5), i.e.

$$\sup_{0 \le r < 1} \int_0^{2\pi} \log^+ \left| F\left(re^{i\varphi}\right) \right| d\varphi = c < \infty.$$

So,

$$\int_{-\infty}^{\infty} \frac{\log^+ |f(x)|}{1 + x^2} \, dx = \frac{1}{2} \int_0^{2\pi} \log^+ \left| f\left(\cot \frac{\theta}{2}\right) \right| d\theta$$

$$= \frac{1}{2} \int_0^{2\pi} \log^+ \left| F\left(e^{-i\theta}\right) \right| d\theta = \frac{1}{2} \int_0^{2\pi} \log^+ \left| F\left(e^{i\varphi}\right) \right| d\varphi.$$

By the Fatou lemma,

$$\int_0^{2\pi} \log^+ \left| F\left(e^{i\varphi}\right) \right| d\varphi \le \sup_{0 \le r < 1} \int_0^{2\pi} \log^+ \left| F\left(re^{i\varphi}\right) \right| d\varphi = c,$$

whence

$$\int_{-\infty}^{\infty} \frac{\log^+ |f(x)|}{1 + x^2} \, dx \le \frac{c}{2} < \infty.$$

Thus, $f(z)$ possesses the property (4.27). Next, taking into account that $f(z)$ is an N-function in both half-planes $\Im z > 0$ and $\Im z < 0$, we find, by Lemma 4.3, that

$$L = \int_{-\infty}^{\infty} \int_{-\infty}^{\infty} \frac{\log^+ |f(z)|}{1 + |z|^4} \, dx \, dy < \infty.$$

Now let the point $\zeta = \xi + i\eta : |\zeta| \ge 1$ be arbitrary. Denote by K_ζ the disk with radius $R = |\zeta|$ and center $z = \zeta$. Since the function $\log^+ |f(z)|$ is subharmonic in K_ζ, the relation

$$\log^+ |f(\zeta)| \le \frac{1}{2\pi} \int_0^{2\pi} \log^+ |f(\xi + \rho \cos \varphi, \eta + \rho \sin \varphi)| \, d\varphi$$

$$= \frac{1}{\pi |\zeta|^2} \iint_{K_\zeta} \log^+ |f(z)| \, dx \, dy \le \frac{1 + 16 |\zeta|^4}{\pi |\zeta|^2} \iint_{K_\zeta} \frac{\log^+ |f(z)|}{1 + |z|^4} \, dx \, dy$$

$$< L \frac{1 + 16 |\zeta|^4}{\pi |\zeta|^2} = \frac{L}{\pi} \left(16 |\zeta|^2 + \frac{1}{|\zeta|^2} \right) \le \frac{17}{\pi} L |\zeta|^2, \quad |\zeta| \ge 1,$$

is valid. Consequently, there exist constants α, $\beta > 0$ such that

$$|f(z)| \leq \alpha e^{\beta |z|^2}. \tag{4.32}$$

Denote by U_δ and V_δ, $0 < \delta < \frac{\pi}{4}$, the angles $\delta \leq \arg z \leq \pi - \delta$ and $|\arg z| \leq \delta$, respectively. According to Theorem 4.4 the growth of the function $f(z)$ is not more than exponential in U_δ. The analogous assertion is valid for the angle U_δ^1, symmetric to U_δ with respect to the origin. As for the angle V_δ, we can apply, taking into account (4.32), the Phragmen-Lindelöf theorem to $f(z)$ (Markushevich [1]), because $2\delta < \frac{\pi}{2}$ and the function $f(z)$ increases not faster than the exponent on the sides of the angle. Thus, the function $f(z)$ is at most of exponential type. A similar conclusion is true for the angle V_δ^1 symmetric to V_δ with respect to the origin. $\qquad\square$

4.5 We shall need some properties of convex domains in $\mathbb{C}^1$. Recall that a nonempty closed set in $\mathbb{C}^1$ is called a convex domain if the fact that two points belong to this set implies that the whole intercept which joins these points belongs to this set too. In particular, any intercept in $\mathbb{C}^1$ is a convex domain. One point is also considered as a convex domain. It is obvious that if an intercept lies completely in a convex domain and has at least one point which is different from its ends at the boundary of this domain, then the whole intercept is a certain part of the boundary.

Let G_1 and G_2 be convex domains in $\mathbb{C}^1$. Then $G_1 + G_2 = \{z : z = z_1 + z_2,\ z_1 \in G_1,\ z_2 \in G_2\}$ is a convex domain. The intersection of any number of convex domains is also a convex one. Thus, the intersection of all convex domains containing a certain set from $\mathbb{C}^1$ is the smallest one which possesses this property. It may also be obtained as the intersection of all the half-planes containing this set. In what follows we deal only with bounded convex domains.

Define the support function of a convex domain $G \subset \mathbb{C}^1$ as

$$k(\theta) = \sup_{z \in G}(x \cos \theta + y \sin \theta) = \sup_{z \in G} \Re\left(z e^{-i\theta}\right), \quad z = x + iy.$$

It follows from the boundedness and closedness of G that the least upper bound in this definition is attained at a certain point from G.

Denote by l_θ the straight line

$$x \cos \theta + y \sin \theta - k(\theta) = 0.$$

It is clear that l_θ has a common point with G. Moreover, since

$$x \cos \theta + y \sin \theta - k(\theta) \leq 0, \quad x + iy \in G,$$

the set G is situated in the same half-plane from the line l_θ. The straight lines l_θ, $\theta \in [0, 2\pi]$, are called supporting for the convex domain G. A point common to

l_θ and G is called the point of support for l_θ. Every supporting straight line has either one point of support or an intercept consisting of points of support.

It is not difficult to understand the geometric meaning of the support function, since $|k(\theta)|$ is the distance from the origin to the supporting straight line l_θ. If the point $z = 0$ belongs to G, then $k(\theta) \geq 0$.

Consider some examples.

The support function of the point $z_0 = \rho e^{i\theta_0}$ coincides with the function

$$k(\theta) = \rho \cos(\theta - \theta_0), \quad \theta \in [0, 2\pi].$$

The functions

$$k(\theta) = H \left|\sin \theta\right|, \quad \theta \in [0, 2\pi],$$

$$k(\theta) = \begin{cases} H \sin \theta & \text{as} \quad 0 \leq \theta \leq \pi \\ 0 & \text{as} \quad \pi < \theta \leq 2\pi, \end{cases}$$

and

$$k(\theta) = \begin{cases} 0 & \text{as} \quad 0 \leq \theta \leq \pi \\ -H \sin \theta & \text{as} \quad \pi < \theta \leq 2\pi \end{cases}$$

are the support functions of the intervals $[-iH, iH]$, $[0, iH]$ and $[-iH, H]$, $H > 0$, respectively. If $G = \{z : |z| \leq R\}$, then

$$k(\theta) \equiv R.$$

Let us note the following properties of a support function:

(i) if $G_1 \subset G$, where G_1 and G are convex domains, and $k_1(\theta)$, $k(\theta)$ their support functions, then

$$k_1(\theta) \leq k(\theta), \quad \theta \in [0, 2\pi];$$

(ii) if the inequality

$$k_1(\theta) < k(\theta), \quad \theta \in [0, 2\pi],$$

holds for the support functions $k_1(\theta)$ and $k(\theta)$ of two convex domains G_1 and G, then $G \supset G_1$ so that every point from G_1 is an interior point of G. The property (i) is evident because the inclusion $G_1 \subset G$ implies

$$k_1(\theta) = \sup_{z \in G_1} \Re(z e^{-i\theta}) \leq \sup_{z \in G} \Re(z e^{-i\theta}) = k(\theta).$$

To prove (ii), suppose that there exists at least one point $z_1 \in G_1$ outside G or at the boundary of G. Then we can draw a straight line l through the point z_1 so that the domain G lies in the same half-plane with respect to l. Taking the ray $\arg z = \alpha$ orthogonal to the line l, we obtain

$$\max_{z \in G} \Re(z e^{-i\alpha}) \leq \max_{z \in G_1} \Re(z e^{-i\alpha}),$$

whence $k(\alpha) \leq k_1(\alpha)$, contrary to the assumption relative to $k(\theta)$ and $k_1(\theta)$.

The conditions below distinguish the support functions of convex domains.

Theorem 4.6 *In order that a function $k(\theta)$, $\theta \in [0, 2\pi]$, be the support function of a convex domain, it is necessary and sufficient that the following conditions be satisfied:*

(i) $k(\theta + 2\pi) = k(\theta)$;

(ii) $\quad k(\theta_1) \sin(\theta_2 - \theta_3) + k(\theta_2) \sin(\theta_3 - \theta_1) + k(\theta_3) \sin(\theta_1 - \theta_2) \leq 0$ (4.33)

for all $\theta_1 \leq \theta_2 \leq \theta_3$, $\theta_3 - \theta_2 < \pi$, $\theta_2 - \theta_1 < \pi$.

Proof. Let G be a convex domain, $k(\theta)$ its support function, and $z = x + iy$ the point on the boundary of G which belongs to the supporting straight line l_{θ_2}. Then

$$x \cos \theta_1 + y \sin \theta_1 - k(\theta_1) \leq 0,$$
$$x \cos \theta_2 + y \sin \theta_2 - k(\theta_2) = 0,$$
$$x \cos \theta_3 + y \sin \theta_3 - k(\theta_3) \leq 0.$$

Multiplying these relations by $\sin(\theta_3 - \theta_2) > 0$, $\sin(\theta_1 - \theta_3)$ and $\sin(\theta_2 - \theta_1) > 0$ respectively and then adding them we obtain the inequality (4.33). The property (i) is evident.

Conversely, suppose the conditions (i) and (ii) to be fulfilled. Choose θ_2 arbitrarily and construct the straight line l_{θ_2}

$$x \cos \theta_2 + y \sin \theta_2 - k(\theta_2) = 0. \tag{4.34}$$

Without loss of generality we can consider $\theta_2 = \frac{\pi}{2}$. The open half-planes

$$x \cos \theta_1 + y \sin \theta_1 - k(\theta_1) > 0, \quad \theta_1 \in (-\frac{\pi}{2}, \frac{\pi}{2}) \tag{4.35}$$

intersect the line $l_{\frac{\pi}{2}}$ by the intervals $(b_1, +\infty)$. The set-theoretic sum of these intervals gives the interval $(b, +\infty)$. Analogously, the union of all the infinite intervals which are the intersections of the half-planes

$$x \cos \theta_3 + y \sin \theta_3 - k(\theta_3) > 0, \quad \theta_3 \in (\frac{\pi}{2}, \frac{3\pi}{2}), \tag{4.36}$$

with $l_{\frac{\pi}{2}}$ yields the interval $(-\infty, a)$. Moreover, no point of the line $l_{\frac{\pi}{2}}$ belongs simultaneously to both intervals $(-\infty, a)$ and $(b, +\infty)$. In the opposite case such a point would satisfy the relations (4.34), (4.35), and (4.36), and, analogous to the proof of the necessity, we should obtain the inequality contrary to (4.33).

Thus $a \leq b$, and every point from the interval $[a, b]$ of the line $l_{\frac{\pi}{2}}$ belongs to all the half-planes

$$x \cos \theta + y \sin \theta - k(\theta) \leq 0.$$

So the intersection of these half-planes is not empty, and it forms a convex domain. Denote it by G. The domain G is situated on the same side with respect to every straight line

$$x \cos \theta + y \sin \theta - k(\theta) = 0$$

and has common points with each of the lines. These straight lines are supporting for the domain G, and $k(\theta)$ is its support function. $\qquad\square$

4.6 Now let $f(z)$ be an entire function at most of exponential type, i.e. the inequality (4.28) is fulfilled. Denote by $h_f(\theta)$ its indicator:

$$h_f(\theta) = \varlimsup_{r \to \infty} \frac{\log |f(re^{i\theta})|}{r}, \quad \theta \in [0, 2\pi]. \tag{4.37}$$

The indicator function $h_f(\theta)$ characterizes dependence of growth of the given function on a direction at which the point z moves to infinity.

One can verify that the function

$$h_g(\theta) = a \cos \theta + b \sin \theta$$

is an indicator of the function

$$g(z) = e^{(a-ib)z}.$$

Such an indicator is called trigonometric.

It is not difficult to make sure that if $h_g(\theta_1) = h_1$ and $h_g(\theta_2) = h_2$ for two points θ_1 and θ_2, then

$$h_g(\theta) = \frac{h_1 \sin(\theta_2 - \theta) + h_2 \sin(\theta - \theta_1)}{\sin(\theta_2 - \theta_1)}. \tag{4.38}$$

Lemma 4.6 *Let $f(z)$ be an entire function of at most exponential type, and $h_f(\theta)$ its indicator. Suppose that*

$$h_f(\theta_1) \le h_1, \ h_f(\theta_2) \le h_2,$$

for some points θ_1 and θ_2 such that $\theta_2 - \theta_1 < \pi$. Then

$$h_f(\theta) \le h_g(\theta), \quad \theta_1 \le \theta \le \theta_2, \tag{4.39}$$

where $h_g(\theta)$ is a trigonometric indicator given by (4.38).

Proof. Let

$$h_g^\delta(\theta) = a_\delta \cos \theta + b_\delta \sin \theta$$

be the function taking the values $h_1 + \delta$ and $h_2 + \delta$, $\delta > 0$, at the points θ_1 and θ_2 respectively. As is easily seen, the indicator of the function

$$\varphi(z) = f(z) e^{-(a_\delta - ib_\delta)z} \quad \text{coincides with} \quad h_\varphi(\theta) = h_f(\theta) - h_g^\delta(\theta).$$

It follows that the function $\varphi(z)$ becomes infinitesimal when z moves along the rays $\arg z = \theta_1$ and $\arg z = \theta_2$. By the Phragmen-Lindelöf theorem, $\varphi(z)$ is bounded in the angle $\theta_1 \le \arg z \le \theta_2$. So, its indicator does not exceed zero. Hence,

$$h_f(\theta) \le h_g^\delta(\theta), \quad \theta_1 \le \theta \le \theta_2$$

for any $\delta > 0$. Passage to the limit as $\delta \to 0$ in this estimate gives the inequality (4.39). $\square$

Lemma 4.6 and Theorem 4.6 lead to the following statement, known as the Polya theorem.

Theorem 4.7 *The indicator $h_f(\theta)$ of an entire function $f(z)$ satisfying the condition (4.28) is the support function $k(\theta)$ of a certain bounded convex domain. This domain is called the indicator diagram of the function $f(z)$.*

Now write the inequality (4.33) for $k(\theta) = h_f(\theta)$ in the form

$$h_f(\theta) \sin(\theta_3 - \theta_1) + h_f(\theta_3) \sin(\theta_1 - \theta) + h_f(\theta_1)\left(\sin(\theta_1 - \theta_3) - \sin(\theta_1 - \theta)\right)$$

$$\leq 4\, h_f(\theta_1) \sin \frac{\theta_3 - \theta_1}{2} \sin \frac{\theta - \theta_1}{2} \sin \frac{\theta_3 - \theta}{2}, \quad \theta_1 \leq \theta \leq \theta_3 .$$

Dividing by $\sin(\theta - \theta_1) \sin(\theta_3 - \theta_1)$ yields

$$\frac{h_f(\theta) - h_f(\theta_1)}{\sin(\theta - \theta_1)}$$

$$\leq \frac{h_f(\theta_3) - h_f(\theta_1)}{\sin(\theta_3 - \theta_1)} + h_f(\theta_1) \sin \frac{\theta_3 - \theta}{2} \sec \frac{\theta_3 - \theta_1}{2} \sec \frac{\theta - \theta_1}{2}. \tag{4.40}$$

Interchanging θ and θ_1, we observe, taking into account the divisor sign, that the inequality (4.40) is also valid for $\theta \leq \theta_1 \leq \theta_3$. If $\theta_3 < \theta_1 < \theta$ or $\theta_3 < \theta < \theta_1$, then the inequality sign in (4.40) is changed. Writing (4.40) for $\theta_1 < \theta < \theta_3$ and $\theta_3 < \theta_1 < \theta$, we establish that the ratio

$$\frac{h_f(\theta) - h_f(\theta_1)}{\sin(\theta - \theta_1)} \tag{4.41}$$

is bounded for $\theta > \theta_1$. The boundedness of the expression (4.41) in the case of $\theta < \theta_1$ is proved in the same way. It follows that the indicator $h_f(\theta)$ is continuous.

We are now in a position to show that the upper limit in (4.37) is attained uniformly, that is for any $\varepsilon > 0$ there exists $R_\varepsilon > 0$ such that

$$\frac{\log |f(r\, e^{i\theta})|}{r} \leq h_f(\theta) + \varepsilon, \quad r > R_\varepsilon , \tag{4.42}$$

for all $\theta \in [0, 2\pi]$.

Indeed, let $\theta_0 = 0, \theta_n = 2\pi, \theta_k (k = 1, \ldots, n - 1)$ be a partition of the interval $[0, 2\pi]$. Let us construct the functions $h_g^j(\theta)$ of the form (4.38), such that

$$h_g^j(\theta_j) = h_f(\theta_j) + \frac{\varepsilon}{3}, \quad h_g^j(\theta_{j+1}) = h_f(\theta_{j+1}) + \frac{\varepsilon}{3} .$$

The subintervals $[\theta_j, \theta_{j+1}]$ may be chosen so small that the oscillation of the functions $h_f(\theta)$ and $h_g^j(\theta)$ is less than $\frac{\varepsilon}{3}$ in each of these intervals. Consequently,

$$h_g^j(\theta) - h_f(\theta) < \frac{2\varepsilon}{3}$$

in every such subinterval.

As was mentioned in the proof of Lemma 4.6, the function

$$\varphi_j(z) = f(z)\, e^{-(a_j - ib_j)z}$$

is bounded in the angle $\theta_j \leq \arg z \leq \theta_{j+1}$. Therefore

$$\log\left|f\left(r\, e^{i\theta}\right)\right| < r\left(h_g^j(\theta) + \frac{\varepsilon}{3}\right)$$

as $r > R_\varepsilon^j$, R_ε^j is sufficiently large. Setting $R_\varepsilon = \max_j R_\varepsilon^j$, we obtain

$$\log\left|f\left(r\, e^{i\theta}\right)\right| < r\left(h_f(\theta) + \varepsilon\right), \quad r > R_\varepsilon,$$

for all $\theta \in [0, 2\pi]$.

By Theorem 4.4, the indicator of an entire function $f(z)$, which is an N-function in both half-planes $\Im z > 0$ and $\Im z < 0$, is determined by the formula (4.21) in the upper half-plane and by a similar one when $\Im z < 0$. Thus,

$$h_f(\theta) = \begin{cases} h_f\left(\frac{\pi}{2}\right)\sin\theta & \text{if} \quad 0 \leq \theta \leq \pi \\ -h_f\left(-\frac{\pi}{2}\right)\sin\theta & \text{if} \quad \pi < \theta \leq 2\pi. \end{cases} \tag{4.43}$$

Hence, the indicator diagram of the function $f(z)$ coincides with the interval of the imaginary axis which joins the points $-ih_f\left(-\frac{\pi}{2}\right)$ and $ih_f\left(\frac{\pi}{2}\right)$. Such functions are of completely regular growth (see Levin [1]), that is the ordinary limit

$$\lim_{r\to\infty} \frac{\log\left|f\left(r\, e^{i\theta}\right)\right|}{r}$$

exists for any $\theta \in [0, 2\pi]$, and

$$\lim_{r\to\infty} \frac{\log\left|f\left(r\, e^{i\theta}\right)\right|}{r} = h_f(\theta), \quad \theta \in [0, 2\pi]. \tag{4.44}$$

The density of zeroes inside any angle $0 < \delta \leq |\arg z| \leq \pi - \delta$ is equal to 0 for such functions. Therefore, its angular density is a step-function which has only two equal jumps at the points 0 and π. The density of zeroes inside the angle $\eta < \theta < \varphi$ is determined by the formula (see Levin [1])

$$\lim_{r\to\infty} \frac{n_f(r, \eta, \varphi)}{r} = \frac{1}{2\pi}\left(h_f'(\varphi) - h_f'(\eta) + \int_\eta^\varphi h_f(\theta)\, d\theta\right), \tag{4.45}$$

where $n_f(r, \eta, \varphi)$ denotes the number of the zeroes of $f(z)$ inside the domain $\{z = |z|\, e^{i\theta} : |z| < r, \eta < \theta < \varphi\}$. If we set

$$n_{f1}(r) = n_f\left(r, -\frac{\pi}{2}, \frac{\pi}{2}\right), \quad n_{f2}(r) = n_f\left(r, \frac{\pi}{2}, \frac{3\pi}{2}\right),$$

we get, due to (4.45),

$$\lim_{r\to\infty} \frac{n_{f1}(r)}{r} = \lim_{r\to\infty} \frac{n_{f2}(r)}{r} = \frac{l}{2\pi}, \tag{4.46}$$

where

$$l = h_f\left(\frac{\pi}{2}\right) + h_f\left(-\frac{\pi}{2}\right).$$

The formula (4.45) also gives for $n_f(r) = n_f(r, 0, 2\pi)$

$$\lim_{r\to\infty} \frac{n_f(r)}{r} = \frac{1}{2\pi} \int_0^{2\pi} h_f(\theta)\, d\theta, \quad \text{whence} \quad \lim_{R\to\infty} \frac{N(R)}{R} = \frac{1}{2\pi} \int_0^{2\pi} h_f(\theta)\, d\theta,$$

where

$$N(R) = \int_0^R \frac{n(r)}{r}\, dr.$$

Using the Jensen formula (4.3), we obtain

$$\lim_{r\to\infty} \int_0^{2\pi} \left(\frac{\log\left|f\left(r\,e^{i\theta}\right)\right|}{r} - h_f(\theta) \right) d\theta = 0. \tag{4.47}$$

Taking into account (4.42), one can conclude from (4.47) and (4.44) that

$$\lim_{r\to\infty} \int_0^{2\pi} \left| \frac{\log\left|f\left(r\,e^{i\theta}\right)\right|}{r} - h_f(\theta) \right| d\theta = 0,$$

which implies the limit equalities

$$\lim_{r\to\infty} \frac{2}{\pi r} \int_0^{\pi} \log\left|f\left(r\,e^{\pm i\theta}\right)\right| \sin\theta\, d\theta$$

$$= \frac{2}{\pi} \int_0^{\pm\pi} h_f(\theta) \sin\theta\, d\theta = \frac{2}{\pi} \int_0^{\pi} h_f\left(\pm\frac{\pi}{2}\right) \sin^2\theta\, d\theta = h_f\left(\pm\frac{\pi}{2}\right). \tag{4.48}$$

As has been mentioned in subsection 4.1, the function $u(z) = \log|f(z)|$ is subharmonic in $\mathbb{R}^2$ if $f(z)$ is an entire function. Applying the analogue of the Phragmen-Lindelöf theorem from that subsection one can obtain, reasoning as when proving Lemma 4.6 and Theorem 4.7, the generalization of the Polya theorem to log-subharmonic functions $u(z)$ (that is non-negative functions for which $\log u(z)$ is subharmonic) satisfying the condition

$$\overline{\lim_{|z|\to\infty}} \frac{\log u(z)}{|z|} < \infty. \tag{4.49}$$

Chapter 2

Entire Operators whose Deficiency Index is (1,1)

Chapter 2 deals mainly with the theory of entire Hermitian operators on a Hilbert space $\mathfrak{H}$, whose deficiency index is (1,1). To characterize the results in detail, recall that the assertions presented in the previous chapter lead to the conclusion that for every simple Hermitian operator A with defect numbers equal to 1 a one-dimensional subspace M (a fixed vector $u \in M : \|u\| = 1$ is called a gauge) can be found so that the original space $\mathfrak{H}$ is decomposed into the direct sum $M \dotplus \mathfrak{M}_z$, $\mathfrak{M}_z = (A - zI)\mathcal{D}(A)$, for each non-real z except for at most a countable set whose limit points may be located only on the real axis. This decomposition generates the map $\Phi : f \mapsto f_u(z)$ from $\mathfrak{H}$ into a certain space of functions $f_u(z)$ meromorphic inside the upper and lower half-planes. At the same time the operator A is transformed into the multiplication by the independent variable. It is established in section 1 that under the condition on the operator A that the set of all f for which the functions $f_u(z)$ are analytic on the whole real axis is dense in $\mathfrak{H}$ (in this case the gauge u is called quasiregular), there exists a bounded non-decreasing function $\sigma(\lambda)$ such that Φ is an isometry from $\mathfrak{H}$ into $L_2(\mathbb{R}^1, d\sigma)$. Such a function $\sigma(\lambda)$ is called a distribution function associated with the gauge u of the operator A. It is not unique if the gauge is fixed. The description of all functions $\sigma(\lambda)$ possessing this property is given in terms of spectral functions of the operator A. Those for which the space $\mathfrak{H}$ is isometrically isomorphic to $L_2(\mathbb{R}^1, d\sigma)$, are distinguished. Though section 2 is auxiliary in character, it is of interest in its own right. It is shown there that if the function $f_u(z)$ is analytic in some neighbourhood $|z - a| < 2r$ of a point $a \in \mathbb{C}^1$, then the inequality $\sup\limits_{z:|z-a|<r} |f_u(z)| \le c\,\|f\|$ holds; moreover, the constant c does not depend on the choice of f. The notion of a u-regular point of the operator A is introduced in section 3. Such a point is understood to be one at which all the functions $f_u(z)$ are analytic. The relation of u-regular points of the operator A to its points of regular type is clarified. In view of the above estimate, $f_u(z)$ is a linear continuous functional on $\mathfrak{H}$ at any u-regular point of A. Therefore it admits a representation of the form $f_u(z) = (f, e(\bar{z}))$. The function $e(z)$ is investigated in section 4. This function characterizes the inclination of the subspace M to $\mathfrak{M}_z$. It plays an important role in problems like the classical moment problem. Section 5 is devoted to study of M-entire operators, that is the operators for which every function $f_u(z)$ is entire. It is proved that in the case of entire operator, the growth of these functions is at most exponential. Their indicator diagrams are certain intervals of the imaginary axis, and the smallest interval on this axis containing them

coincides with the indicator diagram of the log-subharmonic function $\|e(\bar{z})\|$. One of the principal results of this section consists of the fact that the spectrum of any self-adjoint extension within $\mathfrak{H}$ of an entire operator is discrete, and its eigenvalues are zeroes of a corresponding entire function of exponential type which is explicitly expressed through $e(z)$. This made it possible to write the asymptotic distribution formula for the eigenvalues of such extensions. All generalized resolvents of a Hermitian operator with deficiency index $(1,1)$ are described in section 6. The formula obtained there determines a one-to-one correspondence between the set of all the generalized resolvents and the set of all R-functions (i.e. functions analytic in the half-plane $\Im z > 0$ and mapping this half-plane into itself). The central result of the chapter is contained in section 7, where the constructive description of all distribution functions associated with an entire gauge u of an entire operator is given in terms of R-functions, by using an appropriate Nevanlinna matrix. The profound relation of the theory of entire operators to the theory of canonical systems of differential equations is also pointed out there. The directing functionals method is presented in section 8. For a Hermitian operator having one directing functional the spectral theorem on expansion in eigenfunctions of self-adjoint extensions of this operator is established. It should be noted that if a simple Hermitian operator has a so-called universal directing functional, then it is entire. If this is the case, the formula representing $f_u(z)$ by the directing functional is given. The operators entire with respect to a gauge which is a generalized vector (belonging to a certain larger space) are considered in section 9. The majority of the results concerning the theory of entire operators when a gauge is ordinary are extended to the case of a generalized one. The difference between the two cases is that the distribution functions associated with the latter are, generally, not bounded. They may increase; however their growth is at most a power.

1   Quasiregular Gauge

1.1   We define a distribution function as follows.

Definition 1.1 *A bounded function $\sigma(\lambda)$, $\lambda \in \mathbb{R}^1$, is called a distribution function if it possesses the following properties:*

(i) *the function $\sigma(\lambda)$ is non-decreasing;*

(ii) $\sigma(\lambda - 0) = \sigma(\lambda)$;

(iii) $\sigma(-\infty) = 0$.

Denote by $L_2(\mathbb{R}^1, d\sigma)$ the space of all functions $f(\lambda)$ given on $\mathbb{R}^1$ such that

$$\|f\|^2_{L_2(\mathbb{R}^1, d\sigma)} = \int\limits_{-\infty}^{\infty} |f(\lambda)|^2 \, d\sigma(\lambda) < \infty.$$

Let Λ_σ be the operator of multiplication by an independent variable on the space $L_2(\mathbb{R}^1, d\sigma)$:

$$(\Lambda_\sigma f)(\lambda) = \lambda f(\lambda),$$

$$\mathcal{D}(\Lambda_\sigma) = \{f \in L_2(\mathbb{R}^1, d\sigma) : \lambda f \in L_2(\mathbb{R}^1, d\sigma)\}.$$

As is known, the operator Λ_σ is self-adjoint. Its spectral function E_λ is determined by the formula

$$(E_\mu f)(\lambda) = \begin{cases} f(\lambda) & \text{as} \quad \lambda \le \mu \\ 0 & \text{as} \quad \lambda > \mu. \end{cases}$$

Let us return to the isomorphism (1.2.6)

$$f \mapsto f_M(z) = f_u(z)u$$

induced by a simple Hermitian operator A with dense domain. Denote by $\mathfrak{R}$ the set of all the elements $f \in \mathfrak{H}$ whose component $f_M(z)$ is analytic on the whole real axis. By Corollary 1.2.1, for any $f, g \in \mathfrak{R}$,

$$(f, g) = \int_{-\infty}^{\infty} f_u(\lambda)\overline{g_u(\lambda)}\, d\sigma(\lambda),$$

where

$$\sigma(\lambda) = (E_\lambda u, u), \quad \lambda \in \mathbb{R}^1, \tag{1.1}$$

E_λ is a spectral function of the operator A.

Define V_u as the set of all functions of the form (1.1). Every function $\sigma(\lambda) \in V_u$ satisfies the conditions (i), (ii) and (iii) of Definition 1.1. We shall call a function $\sigma(\lambda) \in V_u$ a distribution function associated with the gauge u of the operator A.

Definition 1.2 *We say that the gauge u is quasiregular if the set $\mathfrak{R}$ is dense in $\mathfrak{H}$.*

The following assertion holds for a quasiregular gauge.

Theorem 1.1 *Let the gauge u be quasiregular. In order that a distribution function $\sigma(\lambda)$ belong to the set V_u, it is necessary and sufficient that*

$$(f, g) = \int_{-\infty}^{\infty} f_u(\lambda)\overline{g_u(\lambda)}\, d\sigma(\lambda) \tag{1.2}$$

for any $f, g \in \mathfrak{R}$.

Proof. The necessity of the condition (1.2) has been clarified above. Only the sufficiency remains to be proved.

Assume that the point z_0, $\Im z_0 \neq 0$, is such that

$$\big(u, \varphi(\bar{z}_0)\big) \neq 0. \qquad \text{Then} \qquad \mathfrak{H} = M \dotplus \mathfrak{M}_{z_0}.$$

Denote by A_0 the restriction of the operator A to the set $\mathcal{D}_0 = \mathcal{D}(A) \cap \mathfrak{R}$ and show that $A = \overline{A_0}$. To prove this it suffices to establish that $\mathfrak{M}_{z_0} = \overline{\mathfrak{R}_0}$, where $\mathfrak{R}_0 = \big(A_0 - z_0 I\big)\mathcal{D}_0 \subset \mathfrak{M}_{z_0}$. The latter will be proved if we show that

$$\dim \big(\mathfrak{H} \ominus \mathfrak{R}_0\big) = 1 \;.$$

Suppose that is not the case. Then two orthonormal vectors f_1 and f_2 can be found which are orthogonal to $\mathfrak{R}_0$. The distance from an arbitrary vector f of the form

$$f = c_1 f_1 + c_2 f_2, \quad |c_1|^2 + |c_2|^2 = 1 \,,$$

to $\mathfrak{R}_0$ is equal to 1. Since $\mathfrak{R}$ is dense in $\mathfrak{H}$, we can choose the vectors f_1' and f_2' in $\mathfrak{R}$ so that

$$\|f_j - f_j'\| < \frac{1}{2\sqrt{2}}, \quad j = 1, 2 \,.$$

Then the distance from any vector

$$f' = c_1 f_1' + c_2 f_2', \quad |c_1|^2 + |c_2|^2 = 1 \,,$$

to $\mathfrak{R}_0$ is larger than $\frac{1}{2}$.

On the other hand, we may take the constants c_j, $j = 1, 2$, so that

$$f_M'(z_0) = c_1 f_{1M}'(z_0) + c_2 f_{2M}'(z_0) = 0.$$

By Theorem 1.2.3, the function

$$g_M(z) = \frac{f_M'(z)}{z - z_0}$$

belongs to $\mathfrak{H}_M$ and $f' = \big(A - z_0 I\big)g$, $g \in \mathcal{D}(A)$. As the function $f_M'(z)$ is analytic on the real axis, the function $g_M(z)$ possesses the same property. So $g \in \mathcal{D}(A) \cap \mathfrak{R} = \mathcal{D}_0$ and $f' = \big(A_0 - z_0 I\big)g \in \mathfrak{R}_0$, contrary to the arguments above.

Now let $\sigma(\lambda)$ be a distribution function such that the relation (1.2) holds. The operator

$$\Phi : f \mapsto f_u(\lambda)$$

transforms the set $\mathfrak{R}$ isometrically into the space $L_2\big(\mathbb{R}^1, d\sigma\big)$. Since $\overline{\mathfrak{R}} = \mathfrak{H}$, this operator can be extended to the isometric operator on the whole $\mathfrak{H}$. The operator A_0 is transformed under the map Φ into the restriction of Λ_σ to the set $\Phi\mathcal{D}_0$. Identifying f with Φf, we may consider $\mathfrak{H}$ as a part of $L_2\big(\mathbb{R}^1, d\sigma\big)$. Then the self-adjoint operator Λ_σ is an extension of the operator A_0, hence the operator $A = \overline{A_0}$.

Let E'_λ be the spectral function of Λ_σ. If P is the orthogonal projector from $L_2(\mathbb{R}^1, d\sigma)$ onto $\Phi\mathfrak{H}$, then $E_\lambda = PE'_\lambda$ is a spectral function of the operator A. Therefore

$$
\begin{aligned}
\left(E_\lambda u, u\right) &= \left(\Phi E_\lambda u, \Phi u\right)_{L_2(\mathbb{R}^1, d\sigma)} = \left(\Phi E_\lambda \Phi^{-1} \Phi u, \Phi u\right)_{L_2(\mathbb{R}^1, d\sigma)} \\
&= \left(\Phi E_\lambda \Phi^{-1} 1, 1\right)_{L_2(\mathbb{R}^1, d\sigma)} = \left(PE'_\lambda 1, 1\right)_{L_2(\mathbb{R}^1, d\sigma)} \\
&= \left(E'_\lambda 1, 1\right)_{L_2(\mathbb{R}^1, d\sigma)} = \int_{-\infty}^{\lambda} d\sigma(\mu) = \sigma(\lambda),
\end{aligned}
$$

i.e. the function $\sigma(\lambda)$ is generated by the spectral function E_λ of the operator A and associated with the gauge u. $\qquad\square$

Definition 1.3 *A distribution function $\sigma(\lambda) \in V_u$ is called canonical if it is represented in the form (1.1) where E_λ is an orthogonal spectral function of the operator A.*

Theorem 1.2 *Let the gauge u be quasiregular and $\sigma(\lambda) \in V_u$. The transform $f \mapsto f_u(\lambda)$ from $\mathfrak{R}$ into $L_2(\mathbb{R}^1, d\sigma)$ generates an isometric map Φ from $\mathfrak{H}$ into $L_2(\mathbb{R}^1, d\sigma)$. The latter is an isomorphism from $\mathfrak{H}$ onto $L_2(\mathbb{R}^1, d\sigma)$ if and only if $\sigma(\lambda)$ is a canonical distribution function.*

Proof. The condition is necessary. Indeed, if $\Phi\mathfrak{H} = L_2(\mathbb{R}^1, d\sigma)$, then, after the identification of f with Φf, the operator Λ_σ may be regarded as a self-adjoint extension to $\mathfrak{H}$ of the operator A, and E_λ as an orthogonal spectral function of A. Thus, the distribution function $\sigma(\lambda)$ is canonical.

The proof of sufficiency is more complicated.

Set $\sigma(\lambda) = \left(E_\lambda u, u\right)$, where E_λ is the spectral function of a certain self-adjoint extension $\tilde{A}$ within $\mathfrak{H}$ of the operator A. We must prove the density of the set $\Phi\mathfrak{R}$ in $\mathfrak{H}$, for only in this case $\Phi\mathfrak{H} = L_2(\mathbb{R}^1, d\sigma)$. By (1.2),

$$
\begin{aligned}
\left(\tilde{R}_z f, g\right) &= \int_{-\infty}^{\infty} \frac{d\left(E_\lambda f, g\right)}{\lambda - z} \\
&= \int_{-\infty}^{\infty} \frac{1}{\lambda - z} d \int_{-\infty}^{\lambda} f_u(\mu) \overline{g_u(\mu)} \, d\sigma(\mu) = \int_{-\infty}^{\infty} \frac{f_u(\lambda) \overline{g_u(\lambda)}}{\lambda - z} \, d\sigma(\lambda)
\end{aligned}
\tag{1.3}
$$

for any $f, g \in \mathfrak{H}$, where $\tilde{R}_z = \left(\tilde{A} - zI\right)^{-1}$. Using the Hilbert identity

$$
\tilde{R}_z = \tilde{R}_\zeta + (z - \zeta)\tilde{R}_\zeta \tilde{R}_z,
$$

we get, by virtue of (1.3),

$$
(\widetilde{R}_z f, \widetilde{R}_z f) = (\widetilde{R}_{\bar z} \widetilde{R}_z f, f) = \frac{1}{z - \bar z}\Big((\widetilde{R}_z f, f) - (\widetilde{R}_{\bar z} f, f) \Big)
$$

$$
= \int_{-\infty}^{\infty} \frac{f_u(\lambda)\overline{f_u(\lambda)}}{|\lambda - z|^2}\, d\sigma(\lambda).
$$

(1.4)

In a brief form the formulas (1.3) and (1.4) may be written as

$$
(\widetilde{R}_z f, g) = \Big(\frac{f_u(\lambda)}{\lambda - z}, g_u(\lambda) \Big)_{L_2\left(\mathbb{R}^1, d\sigma\right)},
$$

$$
(\widetilde{R}_z f, \widetilde{R}_z f) = \Big(\frac{f_u(\lambda)}{\lambda - z}, \frac{f_u(\lambda)}{\lambda - z} \Big)_{L_2\left(\mathbb{R}^1, d\sigma\right)}.
$$

It follows from here that

$$
\big\| \widetilde{R}_z f - g \big\|^2 = \Big\| \frac{f_u(\lambda)}{\lambda - z} - g_u(\lambda) \Big\|^2_{L_2\left(\mathbb{R}^1, d\sigma\right)}, \quad f, g \in \mathfrak{R}.
$$

(1.5)

Since $\overline{\mathfrak{R}} = \mathfrak{H}$, for arbitrary $\varepsilon > 0$, $f \in \mathfrak{R}$ and non-real number z, there exists an element $g \in \mathfrak{R}$ such that $\big\| \widetilde{R}_z f - g \big\| < \varepsilon$. In view of (1.5), this means that the closure of the set $\Phi\mathfrak{R}$ in the $L_2\left(\mathbb{R}^1, d\sigma\right)$-norm contains all the functions of the form

$$
\frac{f_u(\lambda)}{\lambda - z}, \quad f_u(\lambda) = \Phi f, \quad f \in \mathfrak{R}.
$$

(1.6)

Now let a function $h(\lambda) \in L_2\left(\mathbb{R}^1, d\sigma\right)$ be orthogonal to $\Phi\mathfrak{R}$. The theorem will be proved if we show that $h(\lambda) = 0$. As $h(\lambda)$ is orthogonal to any function of the form (1.6), the equality

$$
\Big(\frac{f_u(\lambda)}{\lambda - z}, h(\lambda) \Big)_{L_2\left(\mathbb{R}^1, d\sigma\right)} = \int_{-\infty}^{\infty} \frac{f_u(\lambda)\overline{h(\lambda)}}{\lambda - z}\, d\sigma(\lambda) = 0, \quad \Im z \neq 0,
$$

is valid. Taking into account the Stieltjes inversion formula (see Theorem 1.2.4), we conclude that

$$
\int_{\lambda'}^{\lambda''} f_u(\lambda)\overline{h(\lambda)}\, d\sigma(\lambda) = 0
$$

for arbitrary $\lambda' < \lambda''$. The substitution of u for f in this equality leads to

$$
\int_{\lambda'}^{\lambda''} \overline{h(\lambda)}\, d\sigma(\lambda) = 0.
$$

It follows that $h(\lambda) = 0$ σ-almost everywhere, i.e. $h(\lambda)$ is equivalent to zero.   □

2 Principal Theorem on Regular Points of $f_M(z)$

2.1 Let A be a simple Hermitian operator on $\mathfrak{H}$ with dense domain, M a module of the representation (1.2.6) of vectors $f \in \mathfrak{H}$ by their components $f_M(z)$ from M in the decomposition

$$\mathfrak{H} = M \dotplus \mathfrak{M}_z, \quad z \notin S_M, \quad \Im z \neq 0,$$

which are functions meromorphic in the half-planes $\Im z > 0$ and $\Im z < 0$. If u is a gauge of this representation, then $f_M(z) = f_u(z)\, u$ where the function $f_u(z)$ is determined by the formula (1.2.5) as

$$f_u(z) = \frac{\left(f, \varphi(\bar{z})\right)}{\left(u, \varphi(\bar{z})\right)}, \quad z \notin S_M, \quad \Im z \neq 0,$$

$\varphi(z)$ is defined in (1.2.3). Since the function $f_u(z)$ is analytic in the upper and lower half-planes with the exception of the set S_M, for every point $a \notin S_M, \Im a \neq 0$ there exists a disk $K(a,r) = \left\{z : |z - a| < r, \Im z \neq 0\right\}$ in which $\left(u, \varphi(\bar{z})\right) \neq 0$, hence

$$\left|\left(u, \varphi(\bar{z})\right)\right| \geq \mu, \quad z \in K(a,r),$$

where $\mu > 0$ is a constant. Taking into account the analyticity of $\varphi(\bar{z})$ in $K(a,r)$, we obtain

$$\|f_M(z)\| = \frac{\left|\left(f, \varphi(\bar{z})\right)\right|}{\left|\left(u, \varphi(\bar{z})\right)\right|} \|u\| \leq \frac{\|\varphi(\bar{z})\|}{\mu} \|u\| \cdot \|f\| \leq c \|f\| \tag{2.1}$$

as $f \in \mathfrak{H}, z \in K(a,r)$. Here the constant $c = c(a,r) > 0$ does not depend on f.

We shall now show that for any point $a \in \mathbb{C}^1$ (real or non-real) and any number $r > 0$ the constant $c = c(a,r)$ can be found so that the inequality (2.1) is fulfilled if the function $f_M(z)$ is analytic in the disk $K(a, 2r)$. Note at once that $z_0 = i$ and $\varphi_0 : \|\varphi_0\| = \|\varphi(i)\| = 1$ may be taken for the "initial" elements when constructing the function $\varphi(z)$ by the formula (1.2.3). Then

$$\varphi(z) = \varphi(i) + (z - i)\, \mathring{R}_z\, \varphi(i) = \int\limits_{-\infty}^{\infty} \frac{\lambda - i}{\lambda - z} \, dE_\lambda \varphi(i),$$

where E_λ is the spectral function of the extension $\mathring{A}$. Therefore

$$\left(f, \varphi(\bar{z})\right) = \int\limits_{-\infty}^{\infty} \frac{\lambda - i}{\lambda - z} \, d\left(E_\lambda f, \varphi(i)\right) = \int\limits_{-\infty}^{\infty} \frac{\lambda - i}{\lambda - z} \, d\sigma(\lambda, f), \tag{2.2}$$

where

$$\sigma(\lambda, f) = \left(E_\lambda f, \varphi(i)\right).$$

The function $\sigma(\lambda, f)$ satisfies the relation

$$|\sigma(t, f) - \sigma(s, f)| = \left|(E_\Delta f, \varphi(i))\right| = \left|(E_\Delta f, E_\Delta \varphi(i))\right| \leq \left\|E_\Delta f\right\| \left\|E_\Delta \varphi(i)\right\|$$
$$\leq \frac{1}{2} \left(\left\|E_\Delta f\right\|^2 + \left\|E_\Delta \varphi(i)\right\|^2\right) = \frac{1}{2} \left((E_\Delta f, f) + (E_\Delta \varphi(i), \varphi(i))\right)$$

for an arbitrary interval $\Delta = [s, t] \subset \mathbb{R}^1$. So

$$\text{Var } \sigma(\lambda, f) \leq \frac{1}{2} \left(\|f\|^2 + 1\right). \tag{2.3}$$

Thus, for every $f \in \mathfrak{H}$ the function $(f, \varphi(\bar{z}))$ can be represented in the form (1.4.26), where $\sigma(\lambda) = \sigma(\lambda, f)$ is a function of bounded variation. By Lemma 1.4.4, $(f, \varphi(\bar{z}))$ is an N-function in each of the half-planes $\Im z > 0$ and $\Im z < 0$. If $\|f\| = 1$, then, in accordance with (2.3),

$$\text{Var } \sigma(\lambda, f) \leq 1.$$

By Lemma 1.4.5

$$\int\limits_0^\infty \int\limits_{-\infty}^\infty \frac{\log^+ \left|(f, \varphi(\bar{z}))\right|}{(1 + |z|)^4} \, dx \, dy \leq \frac{3}{4}\pi, \quad z = x + iy.$$

A similar inequality may be written for the integral over the lower half-plane. Therefore

$$\int\limits_{-\infty}^\infty \int\limits_{-\infty}^\infty \frac{\log^+ \left|(f, \varphi(\bar{z}))\right|}{(1 + |z|)^4} \, dx \, dy \leq \frac{3}{2}\pi, \quad z = x + iy,$$

if $\|f\| = 1$. By Lemma 1.4.3,

$$\gamma = \int\limits_{-\infty}^\infty \int\limits_{-\infty}^\infty \frac{\log^- \left|(f, \varphi(\bar{z}))\right|}{(1 + |z|)^4} \, dx \, dy < \infty.$$

Since

$$\log^+ \|f_M(z)\| = \log^+ \left|\frac{(f, \varphi(\bar{z}))}{(u, \varphi(\bar{z}))}\right| \|u\|$$
$$\leq \log^+ \left|(f, \varphi(\bar{z}))\right| + \log^- \left|(u, \varphi(\bar{z}))\right| + \log^+ \|u\|,$$

we have

$$\int\limits_{-\infty}^\infty \int\limits_{-\infty}^\infty \frac{\log^+ \|f_M(z)\|}{(1 + |z|)^4} \, dx \, dy \leq \Gamma,$$

where $\Gamma = \frac{3}{2}\pi + \gamma + \frac{\pi}{3} \log^+ \|u\|$. Thus we arrive at the following conclusion.

Lemma 2.1 *A constant* Γ *can be found such that*

$$\int\limits_{-\infty}^{\infty} \int\limits_{-\infty}^{\infty} \frac{\log^+ \|f_M(z)\|}{(1+|z|)^4}\, dx\, dy \leq \Gamma \tag{2.4}$$

for any $f \in \mathfrak{H} : \|f\| = 1.$

It is not difficult now to prove one of the principal theorems of this chapter.

Theorem 2.1 *Let* $a \in \mathbb{C}^1$ *and* $r > 0$ *be arbitrary. There exists a constant* $c = c(a,r)$ *such that the inequality*

$$\|f_M(z)\| \leq c\|f\|, \quad z \in K(a,r) = \{z : |z-a| < r\} \tag{2.5}$$

holds as soon as the function $f_M(z)$ *corresponding to* f *under the isomorphism* (1.2.6) *is analytic in the disk* $K(a,2r)$.

Proof. As the operator $f \mapsto f_M(z)$ is linear, it suffices to make sure that (2.5) is valid in the case of $\|f\| = 1$.

Suppose that the function $f_M(z) = f_u(z)\, u$ associated with $f : \|f\| = 1$ is analytic in $K(a,2r)$. Then the function

$$\log^+ \|f_M(z)\| = \log^+ \left(|f_u(z)|\, \|u\| \right)$$

is subharmonic in $K(a,2r)$, hence

$$\log^+ \|f_M(\zeta)\| \leq \frac{1}{\pi r^2} \iint\limits_{K(\zeta,r)} \log^+ \|f_M(z)\|\, dx\, dy, \quad z = x + iy,$$

for any $\zeta \in K(a,r)$, since $K(\zeta,r) \subset K(a,2r)$.

On the other hand, by (2.4),

$$\iint\limits_{K(\zeta,r)} \log^+ \|f_M(z)\|\, dx\, dy$$

$$\leq (1+|a|+2r)^4 \iint\limits_{K(\zeta,r)} \frac{\log^+ \|f_M(z)\|}{(1+|z|)^4}\, dx\, dy \leq \Gamma\,(1+|a|+2r)^4.$$

So,

$$\|f_M(\zeta)\| \leq \exp\left(\frac{\Gamma(1+|a|+2r)^4}{\pi r^2} \right).$$

This completes the proof. $\qquad\qquad\qquad\qquad\qquad\qquad\qquad\qquad\qquad\qquad\square$

3   *u*-Regular Points of a Hermitian Operator

3.1   Let $A\left(\overline{\mathcal{D}(A)} = \mathfrak{H}\right)$ be a simple closed Hermitian operator on $\mathfrak{H}$ whose defect numbers equal 1, and M and u a module and a gauge respectively of the representation (1.2.6).

Definition 3.1 *We shall call a point $a \in \mathbb{C}^1$ u-regular for the operator A if there exists a disk $K(a, \rho) = \{z : |z - a| < \rho\}$ such that all the functions $f_M(z)$ in the isomorphism (1.2.6) (f goes through the whole space $\mathfrak{H}$) are analytic in this disk.*

Theorem 3.1 *In order that a point $a \in \mathbb{C}^1$ be u-regular for the operator A, it is necessary and sufficient that it be a point of regular type for A and*

$$\mathfrak{M}_a \cap M = 0. \tag{3.1}$$

Proof. If $\Im a \neq 0$, then the first condition is always satisfied. The second one means in this case that

$$\left(u, \varphi(\overline{a})\right) \neq 0, \tag{3.2}$$

where $\varphi(z)$ is given by (1.2.3). According to (1.2.5),

$$f_M(z) = \frac{\left(f, \varphi(\overline{z})\right)}{\left(u, \varphi(\overline{z})\right)} u.$$

So, the function $f_M(z)$ is analytic in every disk $K(a, \rho)$ where $\left(u, \varphi(\overline{z})\right) \neq 0$. If $\left(u, \varphi(\overline{a})\right) = 0$, then the point a is a pole for the functions $g_M(z)$ corresponding to elements g which possess the property $\left(g, \varphi(\overline{a})\right) \neq 0$. Thus, the theorem is true in the case of non-real a.

Suppose $\Im a = 0$. The sufficiency is easily proved in this situation too. Indeed, if a is a point of regular type for the operator A, then, by virtue of Proposition 1.3.1, there exists a self-adjoint extension $\widetilde{A}$ within $\mathfrak{H}$ of the operator A which has no points of spectrum in the disk $K(a, \rho')$ with some $\rho' > 0$. Choose this extension for $\overset{\circ}{A}$ when constructing the vector-function $\varphi(z)$ by the formula (1.2.3). Then the functions $\left(f, \varphi(\overline{z})\right)$ and $\left(u, \varphi(\overline{z})\right)$ are analytic in the disk $K(a, \rho')$, and the representation (1.2.5) holds in $K(a, \rho')$. The conditions (3.1) and (3.2) are equivalent, as before. So, every function $f_M(z)$ is analytic in any disk $K(a, \rho) \subset K(a, \rho')$ in which $\left(u, \varphi(\overline{z})\right) \neq 0$. The verification of the necessity of the condition formulated above in the case of a real a is the subtlest part of the proof. Let us proceed to it.

Assume that a real number a is a u-regular point of the operator A. This means that all the functions $f_M(z)$ are analytic in some disk $K(a, \rho)$. By Theorem 2.1 there exists a constant $c = c\left(a, \frac{\rho}{2}\right)$ such that

$$\left\|f_M(z)\right\| \leq c \|f\|, \quad z \in K\left(a, \frac{\rho}{2}\right), \tag{3.3}$$

for any vector $f \in \mathfrak{H}$. It follows that the functions $f_M(z)$ corresponding to all elements f from the unit ball $S = \{f : \|f\| \leq 1\}$ are bounded uniformly in the disk $K(a, \frac{\varrho}{2})$. According to the Montel compactness principle (see, for instance Markushevich [1]) the set of these functions is compact with respect to the uniform convergence in every interior disk $\{z : |z - a| \leq r\}$, $0 < r < \frac{\varrho}{2}$.

On the other hand, if $\Delta = [a-r, a+r]$, $0 < r < \frac{\varrho}{2}$, then we have, by Corollary 1.2.1,

$$\left(E_\Delta f, g\right) = \int_\Delta f_u(\lambda)\overline{g_u(\lambda)}\, d\sigma(\lambda), \quad f, g \in \mathfrak{H}, \tag{3.4}$$

where $\sigma(\lambda) = \left(E_\lambda u, u\right)$, and E_λ is an arbitrary spectral function of the operator A. Choose from among them an orthogonal one. To prove that a is a point of regular type of the operator A it suffices, in view of Proposition 1.3.3, to show that the spectrum in the interval Δ of the self-adjoint extension $\widetilde{A}$ corresponding to E_λ may consist only of a finite number of eigenvalues of finite multiplicity or that the space $E_\Delta \mathfrak{H}$ is finite-dimensional, which amounts to the same. The latter follows from the fact that the unit ball S_Δ in the space $E_\Delta \mathfrak{H}$ is compact in it. Indeed, if f, g belong to the ball S_Δ, then $(f, g) = \left(E_\Delta f, g\right)$ can be represented in the form

$$(f, g) = \int_\Delta f_u(\lambda)\overline{g_u(\lambda)}\, d\sigma(\lambda).$$

Since the set $\{f_M(z) : f \in S_\Delta \subset S\}$ is compact with respect to the uniform convergence on the interval Δ, the ball S_Δ is also compact in $E_\Delta \mathfrak{H}$.

Now the necessity of the condition (3.1) can be established in the same manner as in the case of $\Im a \neq 0$ if we define the function $\varphi(z)$ using the resolvent $\overset{\circ}{R}_z$ of the extension $\overset{\circ}{A}$ of the operator A which has no points of spectrum in some neighborhood of the point a. This completes the proof. $\square$

The arguments after Proposition 1.3.6 lead to the next conclusion.

Corollary 3.1 *If every point of the real axis is u-regular for the operator A, then the spectrum of any self-adjoint extension within $\mathfrak{H}$ of A is discrete.*

3.2   To be certain that a given point is *u-regular* for the operator A, the following test is useful.

Theorem 3.2 *If for a point $a \in \mathbb{C}^1$, a disk $K(a, \rho)$ and a linear set $\mathfrak{L} : \overline{\mathfrak{L}} = \mathfrak{H}$ can be found so that the function $f_M(z)$ is analytic in $K(a, \rho)$ for each $f \in \mathfrak{L}$, then the point a is u-regular for the operator A.*

Proof. By Theorem 2.1, the inequality (3.3) is satisfied for any $f \in \mathfrak{L}$.

As $\overline{\mathfrak{L}} = \mathfrak{H}$, for every $g \in \mathfrak{H}$ there exists a sequence $\{g_n\}_{n=1}^{\infty} : g_n \in \mathfrak{L}, g_n \to g$ when $n \to \infty$. By virtue of (3.3) the estimate

$$\|g_{nM}(z) - g_{mM}(z)\| \leq c \|g_n - g_m\|, \quad z \in K\left(a, \frac{\rho}{2}\right),$$

holds which implies the uniform convergence inside $K(a, \frac{\rho}{2})$ of the sequence $\{g_{nM}(z)\}_{n=1}^{\infty}$ to a certain function $h(z)$ analytic in $K(a, \frac{\rho}{2})$. Taking into account that

$$g_{nM}(z) = \frac{(g_n, \varphi(\bar{z}))}{(u, \varphi(\bar{z}))} u \to g_M(z), \quad n \to \infty,$$

in any non-real point $z \in K(a, \frac{\rho}{2})$ such that $(u, \varphi(\bar{z})) \neq 0$, we obtain $h(z) = g_M(z)$. So the function $g_M(z)$ is analytic in the disk $K(a, \frac{\rho}{2})$ for an arbitrary $g \in \mathfrak{H}$, which is what had to be proved. □

Theorem 3.3 *A number $a \in \mathbb{R}^1$ is a point of regular type for the operator A if and only if there exist a neighbourhood $K(a, r)$ of this point and an integer $p \geq 0$ such that the function $(z - a)^p f_M(z)$ is analytic in $K(a, r)$ for every $f \in \mathfrak{H}$. In order that a neighborhood $K(a, r)$ and an integer $p \geq 0$ possess these properties, it is sufficient that they be fulfilled for an arbitrary f from a linear set $\mathfrak{L}$ dense in $\mathfrak{H}$.*

Proof. If $a \in \mathbb{R}^1$ is a point of regular type for the operator A, then it is possible to construct the function $\varphi(z)$ in (1.2.3) so that $\varphi(z)$ is analytic in some neighbourhood $K(a, r)$ of a. Because of (1.2.5), this neighborhood and the integer p equal to the multiplicity of a zero of the function $(u, \varphi(\bar{z}))$ at the point $z = a$ are required in the theorem.

Conversely, suppose that a neighborhood $K(a, r)$ of the point a and an integer $p \geq 0$ exist such that the function $(z - a)^p f_M(z)$ is analytic in $K(a, r)$ for each vector f which belongs to a certain linear set $\mathfrak{L}$ dense in $\mathfrak{H}$. We have to prove that a is a point of regular type for the operator A.

The following expansion holds:

$$f_M(z) = \sum_{k=-p}^{\infty} c_k(f)(z - a)^k;$$

here $c_k(f)$ $(k = -p, \ldots, 0, 1, \ldots)$ are linear functionals on $\mathfrak{L}$. Denote by $\mathfrak{L}_0$ the set of all elements $f \in \mathfrak{L}$ for which

$$c_{-1}(f) = c_{-2}(f) = \ldots = c_{-p}(f) = 0. \tag{3.5}$$

The functions $f_M(z)$ corresponding to $f \in \mathfrak{L}_0$ are analytic in $K(a, r)$. It follows from the arguments in the proof of Theorem 3.2 that there exists a neighbourhood $K(a, r_1) \subset K(a, r)$ where all the functions $f_M(z) : f \in \overline{\mathfrak{L}_0}$ are also analytic.

Let P denote the orthogonal projector from $\mathfrak{H}$ onto $\mathfrak{H} \ominus \overline{\mathfrak{L}_0}$. The dimension of the linear set $P\mathfrak{L}$ is not larger than p. Indeed, if $g_i = Pf_i$, $f_i \in \mathfrak{L}$, $i = 1, 2, \ldots, p+1$, then constants $\xi_i : \sum\limits_{i=1}^{p+1} |\xi_i|^2 > 0$ can be selected so that the vector

$$f = \sum_{i=1}^{p+1} \xi_i f_i$$

satisfies the equations (3.5), consequently $f \in \mathfrak{L}_0$. But then

$$Pf = \sum_{i=1}^{p+1} \xi_i g_i = 0.$$

Thus $\dim P\mathfrak{L} \le p$. It follows that

$$\dim P\overline{\mathfrak{L}} = \dim P\mathfrak{H} = \dim\left(\mathfrak{H} \ominus \overline{\mathfrak{L}_0}\right) \le p.$$

Now let $\Delta = (a - h, a + h)$ be an interval lying inside $K(a, r_1)$ and E_λ an orthogonal spectral function of the operator A. If we show that $\dim E_\Delta \mathfrak{H} = \dim\left(E_\Delta \overline{\mathfrak{L}_0} + E_\Delta(\mathfrak{H} \ominus \overline{\mathfrak{L}_0})\right) < \infty$, which is equivalent to $\dim E_\Delta \overline{\mathfrak{L}_0} < \infty$, then the theorem will be proved. But as the equality (3.4) holds for each pair $f, g \in \mathfrak{L}_0$, since the functions $f_M(z)$ and $g_M(z)$ are analytic on Δ, it remains only to repeat the arguments in the proof of Theorem 3.1. □

4 Functions $e(z)$ and $\nabla_M(z)$

4.1   Let z be a u-regular point of a closed simple Hermitian operator on $\mathfrak{H}$ with dense domain. We also suppose its deficiency index to be $(1,1)$. According to Theorem 2.1 the linear operator

$$f \mapsto f_M(z) = f_u(z)\, u, \quad f \in \mathfrak{H},$$

is continuous, hence $f_u(z)$ is a continuous linear functional on $\mathfrak{H}$. By the Riesz theorem it can be represented in the form

$$f_u(z) = (f, e(\bar{z})), \quad f \in \mathfrak{H},$$

where $e(\bar{z})$ is a certain element from $\mathfrak{H}$.

Let a domain $G \subset \mathbb{C}^1$ consist of u-regular points of the operator A. It follows from Definition 3.1 that the function $f_u(z)$ is analytic in G for any $f \in \mathfrak{H}$. Consequently, the vector-function $e(\bar{z})$ is analytic in the mirror image G^* of the set G with respect to the real axis. This property may also be seen directly from the equality

$$e(z) = \frac{\varphi(\bar{z})}{\left(u, \varphi(\bar{z})\right)}, \tag{4.1}$$

which is valid in a sufficiently small neighbourhood of every u-regular point if we take a suitable $\varphi(\bar{z})$.

The formula (4.1) shows that $e(\bar{z}) \in \mathfrak{N}_{\bar{z}}$. So $e(\bar{z})$ is orthogonal to $\mathfrak{M}_z$, whence

$$\left(f - \left(f, e(\bar{z}) \right) u \, , e(\bar{z}) \right) = 0$$

for any $f \in \mathfrak{H}$. From this we get

$$\left(u, \, e(\bar{z}) \right) = 1.$$

This condition together with the inclusion $e(\bar{z}) \in \mathfrak{N}_{\bar{z}}$ determine completely the element $e(\bar{z})$ associated with a u-regular point z of the operator A.

4.2 On the set of all u-regular points of the operator A we define the function $\nabla_M(z)$ as follows:

$$\nabla_M(z) = \sup_{f \in \mathfrak{H}} \frac{\|f_M(z)\|}{\|f\|}. \tag{4.2}$$

The geometric meaning of the value

$$\nabla_M^{-1}(z) = \min_{f \in \mathfrak{H}} \frac{\|f\|}{\|f_M(z)\|}$$

(it will be shown below that the supremum in (4.2) is attained) is sufficiently simple. Namely, the function $f_M(z)$ runs through the whole set M while f goes through the space $\mathfrak{H}$. So, if $g \in M$, then $g = f_M(z)$ for every element $f \in \mathfrak{H}$ of the form $f = g - h$ where $h \in \mathfrak{M}_z$. Thus,

$$\nabla_M^{-1}(z) = \min_{g \in M, h \in \mathfrak{M}_z} \frac{\|g - h\|}{\|g\|} = \min_{g \in M: \|g\|=1, h \in \mathfrak{M}_z} \|g - h\|.$$

As we can see the expression on the right-hand side of the last equality is none other than $\sin \alpha$, where α is the inclination angle of M to the subspace $\mathfrak{M}_z$. The function $\nabla_M(z)$ is log-subharmonic in each domain that consists of u-regular points of A because $\log \nabla_M(z)$ is the upper envelope of the functions $\log \|f_M(z)\|$ ($\|f\| = 1$) subharmonic in this domain (see subsection 1.4.1).

4.3 Suppose that $\|u\| = 1$. Then

$$\|f_M(z)\| = \left| (f, e(\bar{z})) \right|.$$

Writing a vector $f \in \mathfrak{H}$ in the form $f = g + h$ where $g \in \mathfrak{M}_z, h \in \mathfrak{N}_{\bar{z}}$, and taking into account that $\|f_M(z)\| = \|h_M(z)\|$ and $\|h\| < \|f\|$ as $g \neq 0$, we observe that the supremum in (4.2) is attained on the vectors $f \in \mathfrak{N}_{\bar{z}}$, i.e. on the vectors of the form $f = \xi\, e(\bar{z})$, $\xi \in \mathbb{C}^1$. It is easily seen that

$$\nabla_M(z) = \frac{|(e(\bar{z}), e(\bar{z}))|}{\|e(\bar{z})\|} = \|e(\bar{z})\|. \tag{4.3}$$

4.4   Let $\{d_k\}_{k=0}^{\infty}$ be a complete orthonormal system in $\mathfrak{H}$.
Denote by $\{\mathcal{D}_k(z)\, u\}_{k=0}^{\infty}$ its image under the mapping $f \mapsto f_M(z)$. So,

$$\mathcal{D}_k(z) = (d_k, e(\bar{z})).$$

By the Parseval equality

$$\|e(\bar{z})\|^2 = \sum_{k=0}^{\infty} |(e(\bar{z}), d_k)|^2, \quad \text{whence} \quad \sum_{k=0}^{\infty} |\mathcal{D}_k(z)|^2 = \|e(\bar{z})\|^2.$$

Remembering the well-known Dini theorem, we obtain, in view of (4.3) and the continuity of $\mathcal{D}_k(z)$, the following assertion.

Theorem 4.1 *The series*

$$\sum_{k=0}^{\infty} |\mathcal{D}_k(z)|^2$$

converges uniformly on each bounded closed set which consists of u-regular points of the operator A. Its sum does not depend on the choice of the complete orthonormal system $\{d_k\}_{k=0}^{\infty}$, and equals $\nabla_M^2(z)$.

By Theorem 3.2, a point a is u-regular for the operator A if it is possible to find a neighbourhood of this point such that all the vector-functions $\mathcal{D}_k(z)u$, $k \in \mathbb{N}_0 = \{0\} \cup \mathbb{N}$, are analytic there.

5   Entire Operators

In this section we deal again with a closed simple Hermitian operator A whose defect numbers equal 1.

5.1   We start from the definition of an entire operator.

Definition 5.1 *We shall call a closed simple Hermitian operator $A\,(\overline{\mathcal{D}(A)} = \mathfrak{H})$ whose deficiency index is (1,1) entire if the gauge u can be chosen so that the vector-function $f_M(z)$ is entire for every vector $f \in \mathfrak{H}$. If this is the case, then the gauge u is called entire.*

For an entire operator A all points of the complex plane $\mathbb{C}^1$ are u-regular with respect to an entire gauge u. Hence an entire operator is regular. Taking into consideration Theorem 3.1, we may define an entire operator as a regular one for which there exists a module M with the property

$$\mathfrak{M}_z \cap M = 0$$

for all complex z.

If the gauge u is entire, then the function $f_u(z)$ in the equality

$$f_M(z) = f_u(z)\, u$$

is entire. It admits the representation

$$f_u(z) = (f, e(\bar{z}))$$

where the function $e(z)$ taking its values in the space $\mathfrak{H}$, is entire.

As has been clarified in section 2, the function $f_u(z)$ is represented as a quotient of two N-functions in each of the half-planes $\Im z > 0$ and $\Im z < 0$. In the case under consideration, the function $f_u(z)$ being entire is also an N-function in these half-planes. Therefore Theorem 1.4.5 may be applied to this function, and we have

$$\varlimsup_{|z|\to\infty} \frac{\log |(f, e(\bar{z}))|}{|z|} < \infty \tag{5.1}$$

and

$$\int_{-\infty}^{\infty} \frac{\log^+ |(f, e(x))|}{1 + x^2}\, dx < \infty. \tag{5.2}$$

The next lemma is of importance in what follows.

Lemma 5.1 *Let $e(z) \not\equiv 0$ be an entire vector-function with values in $\mathfrak{H}$. Suppose the conditions (5.1) and (5.2) are satisfied for all $f \in \mathfrak{H}$. Then*

$$\varlimsup_{|z|\to\infty} \frac{\log \|e(\bar{z})\|}{|z|} < \infty. \tag{5.3}$$

If we put

$$h_e(\varphi) = \varlimsup_{r\to\infty} \frac{\log \|e(r\, e^{i\varphi})\|}{r}, \qquad -\pi \le \varphi \le \pi, \tag{5.4}$$

then the following equality holds:

$$h_e(\varphi) = \begin{cases} h_e\left(\frac{\pi}{2}\right) \sin\varphi & \text{as} \quad o \le \varphi \le \pi \\ -h_e\left(-\frac{\pi}{2}\right) \sin\varphi & \text{as} \quad -\pi \le \varphi \le 0. \end{cases} \tag{5.5}$$

Proof. As has been shown in Theorem 1.4.4, the function $(f, e(\bar{z}))$ has a continuous indicator

$$h_f(\varphi) = \varlimsup_{r \to \infty} \frac{\log |(f, e(r\,e^{i\varphi}))|}{r}, \quad -\pi \le \varphi \le \pi;$$

moreover, this upper limit is attained uniformly with respect to φ, i.e. for any $\varepsilon > 0$ there exists $R_\varepsilon > 0$ such that

$$\frac{\log |(f, e(r\,e^{i\varphi}))|}{r} \le h_f(\varphi) + \varepsilon, \quad -\pi \le \varphi \le \pi, \quad r > R_\varepsilon. \tag{5.6}$$

The indicator $h_f(\varphi)$ is determined by the formula (1.4.43)

$$h_f(\varphi) = \begin{cases} h_f\left(\frac{\pi}{2}\right) \sin \varphi & \text{as} \quad 0 \le \varphi \le \pi \\ -h_f\left(-\frac{\pi}{2}\right) \sin \varphi & \text{as} \quad -\pi \le \varphi \le 0, \end{cases}$$

and if we set

$$J_n^\pm(f) = \frac{2}{n\pi} \int\limits_0^\pi \log |(f, e(n\,e^{\mp i\varphi}))|\, \sin \varphi \, d\varphi,$$

then, by virtue of (1.4.48),

$$h^\pm(f) = h_f\left(\pm\frac{\pi}{2}\right) = \lim_{n \to \infty} J_n^\pm(f), \quad 0 \ne f \in \mathfrak{H}.$$

The next consideration concerns the functional $h^+(f)$ only; similar arguments are valid for $h^-(f)$. Notice the following properties of the functional $h^+(f)$:

1) $h^+(\lambda f) = h^+(f)$ for any $f \in \mathfrak{H}$, $\lambda \in \mathbb{C}^1$;

2) $h^+(f_1 + f_2) \le \max\{h^+(f_1), h^+(f_2)\}$ for any $f_1, f_2 \in \mathfrak{H}$.

Denote by Ω the set of all elements $f \in \mathfrak{H} : \|f\| \ge 1$. The functionals $J_n^+(f)$, $n \in \mathbb{N}$, are continuous on Ω. This follows from the estimate ($\Omega \ni f_k \to f$ when $k \to \infty$)

$$\left|\log |(f_k, e(n\,e^{-i\varphi}))|\right| |\sin \varphi| \le \log \|f_k\| + \left|\log \|e(n\,e^{-i\varphi})\|\right|$$

$$\le c + \sum_{j=1}^m \rho_j \left|\log |n\,e^{-i\varphi} - a_j|\right| + \left|\log |p(n\,e^{-i\varphi})|\right|$$

and the Lebesgue theorem on passing to the limit under the integral sign; here a_j, $j = 1, \ldots, m$, are the zeroes on the circle $C_n = \{z : |z| = n\}$ of the function $\|e(z)\|$, ρ_j is the multiplicity of a_j, $0 < c = \text{const}$, the function $p(z)$ does not vanish on C_n. As Ω is a complete metric space, the set of all the points of continuity of the functional $h^+(f)$ is, by the Baire-Hausdorff category theorem, dense in Ω. Let $f_0 : \|f_0\| > 1$ be one of them. Then for any $\varepsilon > 0$ there exists $\delta > 0$ such that

$$h^+(f_0 + g) \le h^+(f_0) + \varepsilon \quad \text{as} \quad \|g\| < \delta.$$

So,

$$h^+(g) = h^+\big((f_0 + g) + (g - f_0)\big) \leq \max\big\{h^+(f_0 + g),\, h^+(f_0 - g)\big\} \leq h^+(f_0) + \varepsilon$$

for these ε and g. Due to property 1) of the functional $h^+(f)$ we can omit the condition $\|g\| < \delta$. Since in addition $\varepsilon > 0$ is arbitrary, we get the inequality

$$h^+(f) \leq h^+(f_0)$$

for any $f \in \mathfrak{H}$. Consequently the maxima

$$H^+ = \max_{f \in \mathfrak{H}} h^+(f), \quad H^- = \max_{f \in \mathfrak{H}} h^-(f) \tag{5.7}$$

are finite.

Set

$$H(\varphi) = \begin{cases} H^+ \sin \varphi & \text{as} & 0 \leq \varphi \leq \pi \\ -H^- \sin \varphi & \text{as} & -\pi \leq \varphi \leq 0 \,. \end{cases}$$

According to (5.7),

$$h_f(\varphi) \leq H(\varphi), \quad -\pi \leq \varphi \leq \pi$$

for every $f \in \mathfrak{H}$.

For arbitrary $\varepsilon > 0$, $r > 0$ and $\varphi \in [-\pi, \pi]$, put

$$\Phi_{\varepsilon r \varphi}(f) = \big(f, e(r\, e^{-i\varphi})\big)\, e^{(H(\varphi) + \varepsilon)r}, \quad f \in \mathfrak{H}.$$

Having fixed ε, r and φ we observe that $\Phi_{\varepsilon r \varphi}(f)$ is a continuous linear functional on $\mathfrak{H}$. On the other hand, in view of (5.6), the values of all the functionals $\Phi_{\varepsilon r \varphi}(f)$ at a fixed $f \in \mathfrak{H}$ are bounded if ε is fixed. By the well-known Banach theorem the norms of all $\Phi_{\varepsilon r \varphi}(f)$ are uniformly bounded for every fixed ε, that is, there exists a number $N_\varepsilon > 0$ such that

$$\|\Phi_{\varepsilon r \varphi}(f)\| = \big\|e(r\, e^{-i\varphi})\big\|\, e^{-H(\varphi) + \varepsilon)r} \leq N_\varepsilon,$$

which implies the estimate

$$\big\|e(r\, e^{-i\varphi})\big\| \leq N_\varepsilon\, e^{(H(\varphi) + \varepsilon)r}, \quad r > 0,\ -\pi \leq \varphi \leq \pi\,.$$

Hence, the relation (5.3) is true. We have also proved that

$$h_e(\varphi) \leq H(\varphi), \quad -\pi \leq \varphi \leq \pi\,. \tag{5.8}$$

The arguments at the very end of subsection 1.4.6 and the properties (i) and (ii) of a support function (subsection 1.4.5) show that the indicator diagram of the log-subharmonic function $\|e(\bar{z})\|$ is contained in the interval $[-H^- i,\, H^+ i]$. Therefore it is a certain interval of the imaginary axis. The equality (5.5) follows from here. $\qquad\square$

It is not difficult to see that actually the equality sign holds in (5.8).

Remark 5.1 *The identity*

$$h_e(\varphi) = H(\varphi), \quad -\pi \le \varphi \le \pi, \tag{5.9}$$

is valid, where $h_e(\varphi)$ is defined by (5.4).

Indeed,

$$h_f(\varphi) \le h_e(\varphi) \le H(\varphi), \quad -\pi \le \varphi \le \pi,$$

for any $f \in \mathfrak{H}$. On the other hand, it follows from (5.7) that there exist elements f^+ and f^- such that

$$h^+(f^+) = h_{f+}\left(\frac{\pi}{2}\right) = H\left(\frac{\pi}{2}\right),$$
$$h^-(f^-) = h_{f-}\left(-\frac{\pi}{2}\right) = H\left(-\frac{\pi}{2}\right),$$
whence $\quad h_e\left(\pm\frac{\pi}{2}\right) = H\left(\pm\frac{\pi}{2}\right),$

which implies (5.9).

Lemma 5.1 leads to the following assertion.

Theorem 5.1 *The log-subharmonic function $\nabla_M(z)$ corresponding to an entire gauge $u \in M$ is of at most of exponential type, that is*

$$\varlimsup_{|z| \to \infty} \frac{\log \nabla_M(z)}{|z|} < \infty.$$

Its indicator diagram is a certain interval of the imaginary axis, so that

$$h_{\nabla_M}(\varphi) = \varlimsup_{r \to \infty} \frac{\log \nabla_M(r e^{-i\varphi})}{r}$$
$$= \begin{cases} h_{\nabla_M}\left(\frac{\pi}{2}\right) \sin \varphi & \text{as} \quad 0 \le \varphi \le \pi \\ -h_{\nabla_M}\left(-\frac{\pi}{2}\right) \sin \varphi & \text{as} \quad -\pi \le \varphi \le 0 . \end{cases}$$

This indicator diagram coincides with the smallest interval of the imaginary axis which contains the indicator diagrams of the functions $f_u(z) = (f, e(\bar{z}))$ where f goes through the whole space $\mathfrak{H}$.

The next theorem plays an important role in the investigation of concrete entire operators.

Theorem 5.2 *Let a set $\mathfrak{L}$ be dense in $\mathfrak{H}$. Then the indicator diagram of the function $\nabla_M(z)$ is the smallest interval of the imaginary axis, which contains the indicator diagrams of all the functions $f_u(z)$ corresponding to vectors $f \in \mathfrak{L}$.*

Proof. To prove this statement it is required to show that for an arbitrary $\varepsilon > 0$ we can find elements $g_\pm \in \mathfrak{L}$ such that

$$h_{g+}\left(\frac{\pi}{2}\right) \ge H^+ - \varepsilon, \quad h_{g-}\left(-\frac{\pi}{2}\right) \ge H^- - \varepsilon.$$

Since $H^\pm = h_{f\pm}\left(\pm\frac{\pi}{2}\right)$ (see Lemma 5.1), where $f^\pm$ are points of continuity of the functionals $h_f\left(\frac{\pi}{2}\right)$ and $h_f\left(-\frac{\pi}{2}\right)$ respectively, it is possible to take for $g_\pm$ the elements f from $\mathfrak{L}$ sufficiently close to $f^\pm$. $\qquad\square$

5.2　The length of the indicator diagram of the function $f_u(z)$ does not depend on the choice of the gauge u. In fact, let u and v be two different entire gauges. By virtue of

$$u_v(z) \cdot v_u(z) = \frac{(u, \varphi(\bar{z}))}{(v, \varphi(\bar{z}))} \cdot \frac{(v, \varphi(\bar{z}))}{(u, \varphi(\bar{z}))} = 1,$$

the function $u_v(z)$ vanishes nowhere. Therefore $u_v(z) = e^{g(z)}$, where $g(z)$ is an entire function. It follows from the property (5.1) that $g(z) = \delta z + \gamma$, $\gamma = \text{const}$, $\delta = \text{const}$, i.e.

$$u_v(z) = c e^{\delta z}.$$

As was mentioned in subsection 1.4.6, the indicator of the function

$$u_v(z) = e^{(r_1 \cos \theta + i r_1 \sin \theta) z}, \quad (\delta = r_1 e^{i\theta})$$

is "trigonometric" and is given by the formula

$$h_{u_v}(\varphi) = \varlimsup_{r \to \infty} \frac{\log |u_v(r e^{-i\varphi})|}{r} = r_1 \cos \theta \cos \varphi - r_1 \sin \theta \sin \varphi = r_1 \cos(\theta + \varphi).$$

Since its indicator diagram lies on the imaginary axis, $\delta = i\alpha$, $\Im \alpha = 0$. Remembering that

$$f_u(z) = \frac{(f, \varphi(\bar{z}))}{(u, \varphi(\bar{z}))}, \quad f_v(z) = \frac{(f, \varphi(\bar{z}))}{(v, \varphi(\bar{z}))},$$

we obtain

$$f_v(z) = f_u(z) u_v(z) = c f_u(z) e^{i\alpha z}.$$

Denote by l_v the length of the indicator diagram of the function $f_v(z)$. Then

$$l_v = \varlimsup_{r \to \infty} \frac{\log |f_v(r e^{i\frac{\pi}{2}})|}{r} + \varlimsup_{r \to \infty} \frac{\log |f_v(r e^{-i\frac{\pi}{2}})|}{r}$$

$$= \varlimsup_{r \to \infty} \frac{\log |f_u(r e^{i\frac{\pi}{2}})|}{r} + \varlimsup_{r \to \infty} \frac{\log |f_u(r e^{-i\frac{\pi}{2}})|}{r}$$

$$+ \varlimsup_{r \to \infty} \frac{\log |e^{i\alpha r e^{i\frac{\pi}{2}}}|}{r} + \varlimsup_{r \to \infty} \frac{\log |e^{i\alpha r e^{-i\frac{\pi}{2}}}|}{r} = l_u,$$

which is what had to be proved.　　　　　　　　　　　　　　　　　□

Definition 5.2 *The operator A is called an entire operator of minimal type if the indicator diagram of the function $\nabla M(z)$ consists of zero only. In the opposite case the operator A is said to be of a normal type.*

5.3 Since an entire operator A is regular, the spectrum of its arbitrary self-adjoint extension $\widetilde{A}$ within $\mathfrak{H}$ is discrete. The foregoing results enable us to establish some important characteristics of this spectrum.

Let $\{\varphi_k\}_{k=1}^{\infty}$ be a complete orthonormal system of eigenvectors of the operator $\widetilde{A}$:

$$\widetilde{A}\,\varphi_k \; = \; \lambda_k\,\varphi_k, \quad (\varphi_i, \varphi_k) = \delta_{ik}, \; i, k \in \mathbb{N}.$$

It should be noted that if $\widetilde{A}\,\varphi \; = \; \lambda_0\,\varphi,\; \varphi \neq 0$, then $A^*\,\varphi \; = \; \lambda_0\,\varphi$, i.e. $\varphi \in \mathfrak{N}_{\lambda_0}$, hence

$$\varphi \; = \; c\,e(\lambda_0).$$

Now let ξ be a real number and $\widetilde{A}_\xi$ the self-adjoint extension within $\mathfrak{H}$ of the operator A, for which the number ξ is an eigenvalue (see Proposition 1.3.5). Denote by $e(\xi)$ the corresponding eigenvector.

Consider the function

$$\Phi_\xi(\lambda) \; = \; (\lambda - \xi)\big(e(\lambda), e(\xi)\big).$$

Since the vector $\varphi_k = c\,e(\lambda_k)$ which corresponds to the eigenvalue $\lambda_k \neq \xi$ of the operator $\widetilde{A}_\xi$ is orthogonal to $e(\xi)$, the function $\Phi_\xi(\lambda)$ vanishes at the point λ_k. It is clear that ξ is also a zero of the function $\Phi_\xi(\lambda)$. Thus, all the eigenvalues of the operator $\widetilde{A}_\xi$ are zeroes of the entire functions $\Phi_\xi(\lambda)$.

Conversely, every zero of the function $\Phi_\xi(\lambda)$ is an eigenvalue of the extension $\widetilde{A}_\xi$. To prove this suppose the contrary. Let λ_0 be a zero of the function $\Phi_\xi(\lambda)$ but not an eigenvalue of the operator $\widetilde{A}_\xi$. As the spectrum of $\widetilde{A}_\xi$ is discrete, the point λ_0 is regular for the operator $\widetilde{A}_\xi$. So, the point $z_0 = \lambda_0$ and the resolvent $R_z = (\widetilde{A}_\xi - zI)^{-1}$ can be taken for the "initial" objects φ_0 and $\overset{\circ}{R}_z$ when constructing the vector-function $\varphi(z)$ in (1.2.3):

$$\varphi(z) = e(\lambda_0) + (z - \lambda_0)\,R_z e(\lambda_0).$$

If $\Im z \neq 0$, then

$$\big(\varphi(z), e(\xi)\big) = \big(e(\lambda_0), e(\xi)\big) + \frac{z - \lambda_0}{\xi - z}\,\big(e(\lambda_0), e(\xi)\big) = 0$$

because $\Phi_\xi(\lambda_0) = (\lambda_0 - \xi)\big(e(\lambda_0), e(\xi)\big) = 0$. Since the operator A is simple, the set $\{\varphi(z)\}_{z:\Im z \neq 0}$ is dense in $\mathfrak{H}$, hence $e(\xi) = 0$, which is impossible.

The function $\Phi_\xi(\lambda)$ satisfies the conditions of Theorem 1.4.5. Using that theorem and Theorem 5.1, we conclude that the indicator diagram of the function $\Phi_\xi(\lambda)$ is a certain interval of the imaginary axis whose length does not exceed $h_{\nabla_M}\left(\frac{\pi}{2}\right) + h_{\nabla_M}\left(-\frac{\pi}{2}\right)$. It follows from (1.4.46) that the number $n(r)$ of the eigenvalues of the operator $\widetilde{A}_\xi$ lying in the interval $(-r, r)$ satisfies the relation

$$\lim_{r \to \infty} \frac{n(r)}{r} = \frac{l}{\pi} \leq \frac{1}{\pi}\left(h_{\nabla_M}\left(\frac{\pi}{2}\right) + h_{\nabla_M}\left(\frac{\pi}{2}\right)\right),$$

where l is the length of the indicator diagram of the function $\Phi_\xi(\lambda)$.

Summarizing all that has been said in the present subsection and taking into account Theorems 4.1 and 5.1, we arrive at the following assertion.

Theorem 5.3 *Let A be an entire operator whose deficiency index is (1,1), and $\widetilde{A}_\xi$ its self-adjoint extension within $\mathfrak{H}$ for which the real number ξ is an eigenvalue. The set of the eigenvalues of the operator $\widetilde{A}_\xi$ coincides with the set of the zeroes of the entire function*

$$\Phi_\xi(z) = (z - \xi) \sum_{k=0}^{\infty} \mathcal{D}_k(z)\overline{\mathcal{D}_k(\xi)}. \tag{5.10}$$

Its indicator diagram coincides with a certain interval of the imaginary axis. For different ξ, the zeroes of the functions (5.10) alternate.

6 R-functions and Generalized Resolvents of a Hermitian Operator

Throughout the present section A is assumed to be a closed simple Hermitian operator on $\mathfrak{H}$ with a dense domain and deficiency index (1,1). The relation of its generalized resolvents to a certain class of N-functions is discussed.

6.1 This subsection is devoted to the study of functions analytic in a disk or in a half-plane, taking their values in a certain half-plane. By Theorem 1.4.2 the functions analytic in the unit disk with values on the right half-plane are N-functions. They are distinguished by the following theorem.

Theorem 6.1 *In order that a function $F(\zeta)$ given and finite in the disk $|\zeta| < 1$ be analytic in this disk and $\Re F(\zeta) \geq 0$, it is necessary and sufficient that $F(\zeta)$ admit the representation*

$$F(\zeta) = i\beta + \int_0^{2\pi} \frac{e^{i\theta} + \zeta}{e^{i\theta} - \zeta} d\tau(\theta), \tag{6.1}$$

where $\beta \in \mathbb{R}^1$, $\tau(\theta)$ is a non-decreasing bounded function.

Proof. The sufficiency can be verified directly, for the function in the right-hand side of (6.1) is analytic in the domain $|\zeta| < 1$, and

$$\Re \int_0^{2\pi} \frac{e^{i\theta} + \zeta}{e^{i\theta} - \zeta} d\tau(\theta) = \int_0^{2\pi} \frac{1 - r^2}{1 - 2r\,\cos(\theta - \varphi) + r^2} d\tau(\theta) \geq 0;$$

here $r = |\zeta| < 1$ and $\varphi = \arg \zeta$.

Now let us prove the necessity. Since $F(\zeta)$ is analytic in the disk $|\zeta| < 1$, we have, by the well-known Schwartz formula,

$$F(\zeta) = i\beta + \frac{1}{2\pi} \int_0^{2\pi} \frac{Re^{i\theta} + \zeta}{Re^{i\theta} - \zeta} u(Re^{i\theta}) \, d\theta, \quad |\zeta| \leq R < 1, \tag{6.2}$$

where

$$u(\zeta) = \Re F(\zeta), \quad \beta = \Im F(0).$$

This formula yields

$$\Re F(0) = \frac{1}{2\pi} \int_0^{2\pi} u(Re^{i\theta}) \, d\theta.$$

The representation (6.2) can be rewritten in the form

$$F(\zeta) = i\Im F(0) + \int_0^{2\pi} \frac{Re^{i\theta} + \zeta}{Re^{i\theta} - \zeta} \, d\tau_R(\theta), \tag{6.3}$$

where

$$\tau_R(\theta) = \frac{1}{2\pi} \int_0^{\theta} u(Re^{it}) \, dt.$$

As $u(Re^{it}) \geq 0$, the function $\tau_R(\theta)$ is non-decreasing. Moreover,

$$0 = \tau_R(0) \leq \tau_R(\theta) \leq \tau_R(2\pi) = \Re F(0).$$

Thus, the set of the functions $\tau_R(\theta), 0 < R < 1$, is uniformly bounded with respect to $\theta \in [0, 2\pi]$. By the Helly theorem there exist a non-decreasing function $\tau(\theta), \theta \in [0, 2\pi]$, and a non-decreasing sequence $R_j \to 1, j \to \infty$, such that

$$\lim_{j \to \infty} \tau_{R_j}(\theta) = \tau(\theta).$$

Applying another classical Helly theorem to (6.3) we obtain the representation (6.1). □

Definition 6.1 *A function $f(z)$ analytic in the upper half-plane is called an R-function if its imaginary component is non-negative, i.e.*

$$\Im f(z) \geq 0 \quad as \quad \Im z > 0.$$

The functions

$$f(z) \equiv c, \quad c \in \mathbb{R}^1;$$
$$f(z) = \beta z, \quad 0 < \beta = \text{const};$$
$$f(z) = \frac{1}{t - z}, \quad t \in \mathbb{R}^1,$$

are examples of R-functions. The sum of two R-functions and the multiplication by a positive number of an R-function give R-functions. This and Theorem 1.4.2 show that R-functions form a convex subset in the class of N-functions in the upper half-plane. Note also that no R-function $f(z)$, $\Im z > 0$, except $f(z) \equiv t$, $t \in \mathbb{R}^1$, takes real values. Indeed, if $\Im f(z_0) = 0$ for a point $z_0 : \Im z_0 > 0$, then the function $\Im f(z)$ harmonic in the half-plane $\Im z > 0$ would attain its minimum inside this domain, which is possible only in the case when $\Im f(z) = \text{const}$, $\Im z > 0$. As $\Im f(z_0) = 0$, the function $f(z)$ is a real constant. We also put the function $f(z) \equiv \infty$ into the class of R-functions. So, if $f(z)$, $\Im z > 0$ belongs to the class R-functions, then so does the function $-\frac{1}{f(z)}$. The following criterion gives the description of this class.

Theorem 6.2 *A function $f(z)$ given and finite in the half-plane $\Im z > 0$ is an R-function if and only if it can be represented in the form*

$$f(z) = \alpha + \mu z + \int\limits_{-\infty}^{\infty} \left(\frac{1}{t-z} - \frac{t}{1+t^2} \right) d\sigma(t), \tag{6.4}$$

where $\alpha \in \mathbb{R}^1$, $\mu \geq 0$, $\sigma(t)$ is a non-decreasing function on $\mathbb{R}^1$, such that

$$\int\limits_{-\infty}^{\infty} \frac{d\sigma(t)}{1+t^2} < \infty.$$

If in addition the normalization conditions

$$\sigma(t-0) = \sigma(t), \quad \sigma(0) = 0, \tag{6.5}$$

are fulfilled, then the function $\sigma(t)$ is uniquely determined by $f(z)$.

Proof. If $f(z)$ is of the form (6.4), then

$$\Im f(z) = \mu y + \int\limits_{-\infty}^{\infty} \frac{y}{(t-x)^2 + y^2} \, d\sigma(t) \geq 0 \quad \text{as} \quad y > 0.$$

It is easily seen that $\Im f(z) = 0$ if and only if $\mu = 0$ and $\sigma(t) \equiv \text{const}$, i.e. $f(z) \equiv \alpha$. Suppose now that $f(z)$ is an R-function. The map

$$\zeta = \frac{z-i}{z+i}, \quad z = i\frac{1+\zeta}{1-\zeta}$$

transforms the half-plane $\Im z > 0$ into the disk $|\zeta| < 1$, and the function $f(z)$ into the function

$$F(\zeta) = -if(z).$$

The latter satisfies the conditions of Theorem 6.1. Therefore

$$F(\zeta) = i\beta + \int_0^{2\pi} \frac{e^{i\theta} + \zeta}{e^{i\theta} - \zeta}\, d\tau(\theta) = i\beta + \int_{+0}^{2\pi} \frac{e^{i\theta} + \zeta}{e^{i\theta} - \zeta}\, d\tau(\theta) + \frac{1+\zeta}{1-\zeta}\left(\tau(+0) - \tau(0)\right),$$

whence

$$f(z) = -\beta + \mu z + i\int_{+0}^{2\pi} \frac{e^{i\theta} + \frac{z-i}{z+i}}{e^{i\theta} - \frac{z-i}{z+i}}\, d\tau(\theta) = -\beta + \mu z + \int_{+0}^{2\pi} \frac{iz\frac{e^{i\theta}+1}{e^{i\theta}-1} - 1}{z + i\frac{e^{i\theta}+1}{e^{i\theta}-1}}\, d\tau(\theta)$$

$$= -\beta + \mu z + \int_{+0}^{2\pi} \frac{-z\cot\frac{\theta}{2} + 1}{-\cot\frac{\theta}{2} - z}\, d\tau(\theta) = \alpha + \mu z + \int_{-\infty}^{\infty} \frac{tz+1}{t-z}\, d\omega(t),$$

where

$$\alpha = -\beta, \ \mu = \tau(+0) - \tau(0), \ \int_{-\infty}^{\infty} d\omega(t) = \int_{+0}^{2\pi} d\tau(\theta).$$

Since

$$\frac{1+tz}{t-z} = \left(\frac{1}{t-z} - \frac{t}{1+t^2}\right)(1+t^2),$$

we get

$$f(z) = \alpha + \mu z + \int_{-\infty}^{\infty} \left(\frac{1}{t-z} - \frac{t}{1+t^2}\right) d\sigma(t), \quad \text{where} \quad \sigma(t) = \int_0^t (1+s^2)\, d\omega(s).$$

The estimate

$$\frac{1}{t-z} - \frac{t}{1+t^2} = \frac{1+tz}{t-z} \cdot \frac{1}{1+t^2} = O\left(\frac{1}{1+t^2}\right)$$

implies the absolute convergence of the integral in (6.4). Besides, the constant μ in this representation is determined by the formula

$$\mu = \lim_{y\to\infty} \frac{\Im f(iy)}{y}. \tag{6.6}$$

Indeed,

$$\Im f(iy) = \mu y + \int_{-\infty}^{\infty} \frac{y}{t^2 + y^2}\, d\sigma(t), \quad \text{hence} \quad \frac{\Im f(iy)}{y} = \mu + \int_{-\infty}^{\infty} \frac{1}{t^2 + y^2}\, d\sigma(t).$$

But if $y > 1$, then

$$\left(\int\limits_{-\infty}^{N} + \int\limits_{N}^{\infty} \right) \frac{d\sigma(t)}{t^2 + y^2} < \left(\int\limits_{-\infty}^{N} + \int\limits_{N}^{\infty} \right) \frac{d\sigma(t)}{t^2 + 1} < \frac{\varepsilon}{2}$$

for sufficiently large N. At the same time

$$\int\limits_{-N}^{N} \frac{d\sigma(t)}{t^2 + y^2} < \frac{\varepsilon}{2}$$

if y is large enough. Consequently the equality (6.6) is true.

The uniqueness of the function $\sigma(t)$ satisfying the normalization conditions (6.5) is established by using Theorem 1.2.4. In the case under consideration

$$\Im f(x + iy) - \mu y = \int\limits_{-\infty}^{\infty} \frac{y}{(t - x)^2 + y^2}\, d\sigma(t).$$

Since the function $\sigma(t)$ is real, the Stieltjes inversion formula (1.2.9), where

$$\varphi(\lambda) \equiv 1, \quad F(z) = \int\limits_{-\infty}^{\infty} \frac{d\sigma(t)}{t - z},$$

$a, b \in \mathbb{R}^1$ are arbitrary points of continuity of $\sigma(t)$, takes the form

$$\sigma(b) - \sigma(a) = \lim_{y \to 0} \frac{1}{\pi} \int\limits_{a}^{b} \Im F(x + iy)\, dx = \lim_{y \to 0} \frac{1}{\pi} \int\limits_{a}^{b} \frac{y}{(t - x)^2 + y^2}\, d\sigma(t). \qquad \square$$

Theorem 6.3 *In order that a function $f(z)$, $\Im z > 0$, admit a representation of the form*

$$f(z) = \gamma + \int\limits_{-\infty}^{\infty} \frac{d\sigma(t)}{t - z}, \tag{6.7}$$

where $\gamma \in \mathbb{R}^1$, $\sigma(t)$ is a bounded non-decreasing function, it is necessary and sufficient that $f(z)$ be an R-function and

$$y\, \Im f(iy) \le c, \quad 0 < c = \text{const}, \quad y > 0.$$

If the function $\sigma(t)$ obeys the normalization conditions (6.5), then $\sigma(t)$ is uniquely determined by $f(z)$.

Proof. It is evident that a function of the form (6.7) is an R-function. Moreover,

$$y \, \Im f(iy) = \int_{-\infty}^{\infty} \frac{y^2}{t^2 + y^2} \, d\sigma(t) \le \int_{-\infty}^{\infty} d\sigma(t) = c = \text{const}.$$

Conversely, the function $f(z)$ being an R-function can be represented in the form (6.4). In view of the inequality $y \, \Im f(iy) \le c$, the estimate

$$y \, \Im f(iy) = \mu y^2 + \int_{-\infty}^{\infty} \frac{y^2}{t^2 + y^2} \, d\sigma(t) \le c$$

holds, whence $\mu y^2 \le c$. As y is arbitrary, $\mu = 0$. Next,

$$c \ge y \, \Im f(iy) \ge \int_{-N}^{N} \frac{y^2}{t^2 + y^2} \, d\sigma(t).$$

Passing to the limit in this inequality when $y \to \infty$, we obtain

$$-\int_{-N}^{N} d\sigma(t) \le c \quad \text{for any } N > 0. \text{ Therefore} \quad \int_{-\infty}^{\infty} d\sigma(t) \le c.$$

Now the difference under the integral sign in (6.4) can be integrated termwise, and we may add the integral

$$\int_{-\infty}^{\infty} \frac{t}{1 + t^2} \, d\sigma(t)$$

to the constant α. The uniqueness of the measure $d\sigma(t)$ under the normalization conditions follows from the previous theorem.   □

It has been seen in the course of proving the theorem that

$$\sup_{0 < y < \infty} y \, \Im f(iy) = \lim_{y \to \infty} y \, \Im f(iy).$$

Definition 6.2 *We call a function $f(z)$ given in the half-plane $\Im z > 0$ an R_0-function if it admits a representation of the form*

$$f(z) = \int_{-\infty}^{\infty} \frac{d\sigma(t)}{t - z}, \quad \int_{-\infty}^{\infty} d\sigma(t) < \infty, \tag{6.8}$$

where $\sigma(t)$ is a non-decreasing function.

Theorem 6.4 *A function $f(z)$, $\Im z > 0$ is an R_0-function if and only if the following conditions are satisfied:*

1) $f(z)$ *is an R-function;*

2) $y \Im f(iy) \leq c$, $y > 0$;

3) $\lim\limits_{y \to \infty} f(iy) = 0$.

The function $\sigma(t)$ normed by the conditions (6.5) is uniquely determined by $f(z)$.

Proof. We must prove only that in the representation (6.7)

$$\gamma = \lim_{y \to \infty} f(iy). \tag{6.9}$$

But

$$\left| \int_{-\infty}^{\infty} \frac{d\sigma(t)}{t - iy} \right| \leq \int_{-\infty}^{\infty} \frac{d\sigma(t)}{\sqrt{t^2 + y^2}} = \left(\int_{-\infty}^{N} + \int_{N}^{\infty} \right) \frac{d\sigma(t)}{\sqrt{t^2 + y^2}} + \int_{-N}^{N} \frac{d\sigma(t)}{\sqrt{t^2 + y^2}}.$$

Both terms in the right-hand side can be made as small as desired by choosing N large enough in the first one and y sufficiently large in the second one. Consequently (6.9), hence (6.8) is valid.   □

6.2   In what follows attention is focused on the study of generalized resolvents of the operator A.

Let $\tilde{A}$ be a self-adjoint extension, in general, with exit to a larger space $\tilde{\mathfrak{H}} \supseteq \mathfrak{H}$, of the operator A, $\tilde{R}_z$ its resolvent.

Definition 6.3 *The operator-function*

$$R_z = P\tilde{R}_z$$

given on the set of regular points of $\tilde{A}$ where P is the orthoprojector from $\tilde{\mathfrak{H}}$ onto $\mathfrak{H}$, is called the generalized resolvent of A, generated by the extension $\tilde{A}$. The resolvents generated by self-adjoint extensions within $\mathfrak{H}$ are called canonical or orthogonal ones.

As before we put

$$\mathfrak{M}_z = (A - zI)\mathcal{D}(A), \quad \mathfrak{N}_z = \mathfrak{H} \ominus \mathfrak{M}_{\bar{z}}, \quad \Im z \neq 0.$$

Fix now a self-adjoint extension $\overset{\circ}{A}$ within $\mathfrak{H}$ of the operator A and set $\overset{\circ}{R}_z = \left(\overset{\circ}{A} - zI\right)^{-1}$. If $f \in \mathfrak{M}_z$, then $f = (A - zI)g$, $g \in \mathcal{D}(A)$, and

$$\left(\overset{\circ}{R}_z - R_z\right)f = \left((\overset{\circ}{A} - zI)^{-1}(A - zI)g - P(\tilde{A} - zI)^{-1}(A - zI)g = g - g = 0.$$

Since the operators on $\mathfrak{H}$

$$\overset{\circ}{U}_{\bar{z}z} = I + (z - \bar{z})\overset{\circ}{R}_z$$

and

$$U_{\bar{z}z} = P\tilde{U}_{\bar{z}z} = P\big(I + (z - \bar{z})\tilde{R}_z\big) = I + (z - \bar{z})R_z \tag{6.10}$$

map $\mathfrak{N}_{\bar{z}}$ into $\mathfrak{N}_z$, we have for the difference of the resolvents $\overset{\circ}{R}_z$ and R_z

$$\begin{cases} \big(R_z - \overset{\circ}{R}_z\big)f = 0 & \text{as} \quad f \in \mathfrak{M}_z \\ \big(R_z - \overset{\circ}{R}_z\big)f \in \mathfrak{N}_z & \text{as} \quad f \in \mathfrak{N}_{\bar{z}}. \end{cases} \tag{6.11}$$

As we can see, the operator $R_z - \overset{\circ}{R}_z$ maps the whole space $\mathfrak{H}$ into $\mathfrak{N}_z$. In view of $\dim \mathfrak{N}_z = 1$, we get

$$\big(R_z - \overset{\circ}{R}_z\big) = c_{fz}\varphi(z), \tag{6.12}$$

where

$$\varphi(z) = \varphi(z_0) + (z - z_0)\overset{\circ}{R}_z\varphi(z_0) \in \mathfrak{N}_z, \tag{6.13}$$

$z_0 : \Im z_0 \neq 0$ is fixed, $\|\varphi(z_0)\| = 1$, and the constant c_{fz} depends on f, z and $\tilde{A}$.

The equality (6.12) shows that c_{fz} is a linear continuous functional on $\mathfrak{H}$, and therefore

$$c_{fz} = \big(f, h(z)\big).$$

It follows from (6.11) that $\big(f, h(z)\big) = 0$ for any f orthogonal to $\mathfrak{N}_{\bar{z}}$, i.e. $h(z) = \overline{\gamma(z)}\varphi(\bar{z})$. The function $\gamma(z)$ is analytic at the common points of analyticity of the operator-functions R_z and $\overset{\circ}{R}_z$. The equality (6.12) can now be written in the form

$$R_z f = \overset{\circ}{R}_z f + \gamma(z)\big(f, \varphi(\bar{z})\big)\varphi(z). \tag{6.14}$$

Define the function $Q(z)$ as

$$Q(z) = \gamma^{-1}(z).$$

Suppose first that $\tilde{A}$ is a self-adjoint extension within $\mathfrak{H}$ of the operator A. Substituting R_z and R_{z_0} expressed in the form (6.14) into the Hilbert resolvent identity

$$R_z = R_{z_0} + (z - z_0)R_z R_{z_0},$$

we obtain, by using this identity applied to $\overset{\circ}{R}_z$ and then the relation (6.13), that

$$Q(z) = Q(z_0) + (z - z_0)\big(\varphi(z_0), \varphi(\bar{z})\big). \tag{6.15}$$

It should be noted that when $\tilde{A}$ is an extension with exit to a larger space $\tilde{\mathfrak{H}} \supset \mathfrak{H}$, the equality (6.15) for the function $Q(z)$ corresponding to $\tilde{A}$ is generally not true.

6.3   Let us consider the nature of the function $\gamma(z)$. First find $\gamma(z_0)$. For this purpose substitute z_0 for z in (6.14) and pass from the resolvents $\overset{\circ}{R}_z$ and R_z to the Caley transforms $\overset{\circ}{\tilde{U}}_{\bar{z}z}$ and $\tilde{U}_{\bar{z}z}$. Then (6.10) and (6.14) imply

$$U_{\bar{z}_0 z_0} f = P\tilde{U}_{\bar{z}_0 z_0} f = \overset{\circ}{\tilde{U}}_{\bar{z}_0 z_0} f + (z_0 - \overline{z_0})\gamma(z_0)\big(f, \varphi(\overline{z_0})\big)\varphi(z_0)\,.$$

Setting $f = \varphi(\overline{z}_0)$, we arrive at the equality

$$U_{\bar{z}_0 z_0}\varphi(\overline{z}_0) = \varphi(z_0) + (z_0 - \overline{z_0})\gamma(z_0)\varphi(z_0)\,. \tag{6.16}$$

Since

$$U_{\bar{z}_0 z_0}\varphi(\overline{z}_0) \in \mathfrak{N}_{z_0}\,,$$

we have

$$U_{\bar{z}_0 z_0}\varphi(\overline{z}_0) = \theta\varphi(z_0)\,; \tag{6.17}$$

here $\theta = \theta(z_0)$ is a parameter which is determined by the extension $\tilde{A}$. Its module does not exceed 1 and is equal to 1 if and only if $\tilde{A}$ is an extension within $\mathfrak{H}$.

The above arguments lead to the following assertion.

Proposition 6.1 *If the equality*

$$R_z = \overset{\circ}{R}_z\,,$$

or even the partial one

$$R_z\varphi(\overline{z}) = \overset{\circ}{R}_z\varphi(\overline{z})\,,$$

is satisfied for at least one non-real point $z = z_1$, then it holds for any $z : \Im z \neq 0$.

Proof. The formula (6.14) yields $\gamma(z_1) = 0$. Taking $z_0 = z_1$ in (6.16) and (6.17), we infer that $\theta = \theta(z_1) = 1$. So $\tilde{A}$ is an extension within $\mathfrak{H}$ of the operator A. But the resolvents of two self-adjoint extensions within $\mathfrak{H}$ of the same Hermitian operator, which coincide at one point, are evidently identical.   □

As we can observe in the proof of the proposition, $\gamma(z) \neq 0$ if $\Im z \neq 0$. Comparing (6.16) and (6.17) gives

$$1 + (z_0 - \overline{z}_0)\gamma(z_0) = \theta\,,$$

whence

$$\gamma(z_0) = \frac{1 - \theta}{z_0 - \overline{z}_0}\,, \qquad \text{or} \qquad Q(z_0) = \gamma^{-1}(z_0) = i\Im z_0 + \tau\,,$$

where the parameter τ related to θ by

$$\tau = i\,\frac{1 + \theta}{1 - \theta}\,\Im z_0$$

maps the unit disk of θ-plane onto the upper half-plane of the τ-plane.

If R_z is the resolvent of a self-adjoint extension within $\mathfrak{H}$, then the equality (6.15) is true, and the formula (6.14) takes the form

$$R_z f = \overset{\circ}{R}_z f + \frac{(f, \varphi(\bar{z}))\varphi(z)}{\tau + Q_1(z)}, \quad \Im \tau = 0, \tag{6.18}$$

where

$$Q_1(z) = i\,\Im z_0 + (z - z_0)(\varphi(z_0), \varphi(\bar{z})). \tag{6.19}$$

When the parameter τ ranges over the whole real axis, R_z passes through the set of the resolvents of all self-adjoint extensions within $\mathfrak{H}$ of the operator A. The correspondence defined by (6.18) between all the extensions within $\mathfrak{H}$ and the values of the parameter $\tau \in \mathbb{R}^1$ is one-to-one.

Let us now concentrate on the study of properties of the function $Q_1(z)$. Put $z_0 = \zeta$ in (6.15) and (6.19). Then

$$\frac{Q_1(z) - Q_1(\zeta)}{z - \zeta} = \frac{Q(z) - Q(\zeta)}{z - \zeta} = (\varphi(\zeta), \varphi(\bar{z})).$$

The interchange of z and ζ leads to the relation

$$(\varphi(z), \varphi(\bar{\zeta})) = (\varphi(\zeta), \varphi(\bar{z})).$$

Taking $\bar{z}$ instead of ζ, we get

$$\frac{Q_1(z) - Q_1(\bar{z})}{z - \bar{z}} = (\varphi(\bar{z}), \varphi(\bar{z})) \geq 0,$$

which shows that $\Re Q_1(z) = \Re \overline{Q_1(\bar{z})}$. Since the functions $Q_1(z)$ and $\overline{Q_1(\bar{z})}$, $\Im z \neq 0$, are analytic, they differ from each other by a constant. Taking into account that due to (6.19)

$$Q_1(z_0) = i\,\Im z_0, \quad Q_1(\bar{z}_0) = -i\,\Im z_0,$$

we conclude that the function $Q_1(z)$ is real. Moreover, (6.19) yields $Q'_1(z_0) = (\varphi(z_0), \varphi(\bar{z}_0))$, and if $x \in \mathbb{R}^1$ is a regular point of the operator A, then

$$Q'_1(x) = (\varphi(x), \varphi(x)) > 0.$$

Consequently, the function $Q_1(x)$ is monotone increasing.

In the general case when R_z is the resolvent of a self-adjoint extension with exit into a larger space $\widetilde{\mathfrak{H}} \supset \mathfrak{H}$ of the operator A, we define $\gamma(z)$ in the formula (6.14) as

$$\gamma(z) = Q^{-1}(z) = (\tau(z) + Q_1(z))^{-1}.$$

Thus,

$$R_z f = \overset{\circ}{R}_z f + \frac{(f, \varphi(\bar{z}))\varphi(z)}{\tau(z) + Q_1(z)}, \tag{6.20}$$

where $\tau(z)$ is a function analytic in the upper half-plane whose imaginary component is non-negative there. Since $\Im \tau(z_0) \geq 0$, to prove the last statement it suffices to show that the equality $\Im\tau(z') = 0$ at least for one point $z' : \Im z' > 0$ implies $\tau(z) \equiv \text{const}$ for $z : \Im z > 0$. It will follow from here that the inequality $\Im\tau(z) < 0$ is not possible at any point z of the upper half-plane. So, suppose $\Im\tau(z') = 0 \, (\Im z' > 0)$. Put $z = z'$ in (6.18) and (6.20) and choose $\tau(z')$ as the constant τ in (6.18). The right-hand sides of (6.18) and (6.20) are identical for every $f \in \mathfrak{H}$. Hence, the generalized resolvent R_z coincides at the point z' with the resolvent of a certain self-adjoint extension within $\mathfrak{H}$ of the operator A. By Proposition 6.1, these resolvents are identical. Therefore $\tau(z) = \tau(z')$ if $\Im z \neq 0$.

6.4   Multiplying the two sides of (6.20) scalarly by f and then reducing to a common denominator, we arrive at the representation

$$\left(R_z f, f\right) = \frac{\gamma_0(z) + \gamma_1(z)\tau(z)}{Q_1(z) + \tau(z)}, \quad \Im z > 0,$$

where for a fixed $f \in \mathfrak{H}$ the functions $\gamma_0(z), \gamma_1(z), Q_1(z)$ do not depend on the choice of a resolvent determining the function $\tau(z)$. Since the canonical resolvents correspond to $\tau(z) \equiv \tau \in \mathbb{R}^1$ and the rest to $\tau(z)$ running a certain set of analytic functions whose imaginary components are non-negative, the point

$$w = \left(R_z f, f\right)$$

falls into a certain circular domain. Its boundary, denoted by $C(f, z)$, is a circle which is run by the point w when R_z passes through all the canonical resolvents of A. In other words, $C(f, z)$ is circumscribed by the point

$$w = \frac{\gamma_0(z) + \gamma_1(z)\tau}{Q_1(z) + \tau}$$

when τ goes through the whole real axis. Since the set of all generalized resolvents of A is convex, the points $\left(R_z f, f\right)$ fill the whole disk $K(f, z)$ inside the circle $C(f, z)$ when R_z runs through the set of values at the point z of all operator-functions R_z.

The foregoing consideration results in the following theorem.

Theorem 6.5 *Any generalized resolvent R_z of the operator A can be represented in the form*

$$R_z = \overset{\circ}{R}_z + \frac{(\cdot, \varphi(\bar{z}))\varphi(z)}{\tau(z) + Q_1(z)}, \tag{6.21}$$

where $\tau(z)$ is an R-function. Conversely, every R-function $\tau(z)$ generates a certain generalized resolvent of the operator A by the formula (6.21). The canonical resolvents of A are given by (6.21) where $\tau(z) \equiv \tau = \text{const}, \tau \in \mathbb{R}^1$.

Proof. We must prove only the converse statement.

Suppose $\tau(z)$ to be an R-function. Denote by S_z the operator on $\mathfrak{H}$ defined by the expression

$$S_z = \overset{\circ}{R}_z + \frac{(\cdot, \varphi(\bar{z}))\varphi(z)}{\tau(z) + Q_1(z)}. \tag{6.22}$$

The function $(S_z f, f)$ is analytic in the half-plane $\Im z > 0$ for any $f \in \mathfrak{H}$. Since the circle $C(f, z)$ is situated on the upper half-plane and the point $(S_z f, f)$ lies on the disk $K(f, z)$, this function belongs to the class of R-functions. In view of the estimate

$$|(R_z f, f)| \le \frac{(f, f)}{\Im z}, \quad \Im z > 0,$$

valid for canonical resolvents of the operator A, the value $(S_z f, f)$ belonging to the disk $K(f, z)$ satisfies the inequality

$$|(S_z f, f)| \le \frac{(f, f)}{\Im z}, \quad \Im z > 0,$$

whence

$$y|(S_z f, f)| \le (f, f), \quad y > 0. \tag{6.23}$$

Thus, $(S_z f, f)$ is an R_0-function. By Theorem 6.4,

$$(S_z f, f) = \int_{-\infty}^{\infty} \frac{d\sigma(\lambda, f)}{\lambda - z}, \quad \Im z > 0, \tag{6.24}$$

where the non-decreasing function $\sigma(\lambda, f) = \sigma(\lambda - 0, f)$, $\sigma(-\infty) = 0$, is uniquely determined by $(S_z f, f)$. By virtue of the identity $S_{\bar{z}} = S_z^*$, the representation (6.24) can be extended to all $z : \Im z \ne 0$.

Set

$$\sigma(\lambda, f, g) = \frac{1}{4}\sigma(\lambda, f + g) - \frac{1}{4}\sigma(\lambda, f - g) + \frac{i}{4}\sigma(\lambda, f + ig) - \frac{i}{4}\sigma(\lambda, f - ig).$$

The function $\sigma(\lambda, f, g)$, $\lambda \in \mathbb{R}^1$, has a bounded variation. It is left-continuous at every point $\lambda > -\infty$ and $\sigma(\lambda, f, g) \to 0$ as $\lambda \to -\infty$. It follows from (6.24) that

$$(S_z f, g) = \int_{-\infty}^{\infty} \frac{d\sigma(\lambda, f, g)}{\lambda - z}. \tag{6.25}$$

The representations (6.25) and (6.24) are unique. This and the identity $S_{\bar{z}} = S_z^*$ yield

$$\overline{\sigma(\lambda, f, g)} = \sigma(\lambda, g, f) \tag{6.26}$$

and
$$\sigma(\lambda, \alpha_1 f_1 + \alpha_2 f_2, g) = \alpha_1 \sigma(\lambda, f_1, g) + \alpha_2 \sigma(\lambda, f_2, g).$$

If we show that
$$\sigma(\lambda, f) = \sigma(\lambda, f, f) \le (f, f), \quad f \in \mathfrak{H}, \tag{6.27}$$

then $\sigma(\lambda, f, g)$ will be a bilinear functional on $\mathfrak{H}$ whose norm does not exceed 1.
Due to (6.23),
$$\left| \int_{-\infty}^{\infty} \frac{y\, d\sigma(\lambda, f)}{\lambda - iy} \right| \le (f, f),$$

whence
$$\left| \int_{-a}^{a} \frac{y\, d\sigma(\lambda, f)}{\lambda - iy} \right| \le (f, f) + \int_{-\infty}^{-a} d\sigma(\lambda, f) + \int_{a}^{\infty} d\sigma(\lambda, f)$$

for any $a > 0$. Passing in this inequality to the limit as $y \to \infty$, we obtain
$$\int_{-a}^{a} d\sigma(\lambda, f) \le (f, f) + \int_{-\infty}^{-a} d\sigma(\lambda, f) + \int_{a}^{\infty} d\sigma(\lambda, f),$$

which gives, by passing to the limit as $a \to \infty$,
$$\sigma(\infty, f, f) \le (f, f).$$

Taking into account that the function $\sigma(\lambda, f)$ is non-decreasing, we conclude that the estimate (6.27) is true. So, there exists a family of operators F_λ, $\lambda \in \mathbb{R}^1$, such that
$$\sigma(\lambda, f, g) = (F_\lambda f, g), \quad f, g \in \mathfrak{H},$$

hence
$$(S_z f, g) = \int_{-\infty}^{\infty} \frac{d(F_\lambda f, g)}{\lambda - z}. \tag{6.28}$$

The self-adjointness of F_λ is the result of (6.26). In addition, the family F_λ, $\lambda \in \mathbb{R}^1$, is uniformly bounded:
$$(F_\lambda f, f) \le (f, f), \quad \lambda \in \mathbb{R}^1. \tag{6.29}$$

The normalization conditions for the function $\sigma(\lambda)$ give, first in the weak and then in the strong sense,
$$F_{-\infty} = 0, \quad F_{\lambda - 0} = F_\lambda.$$

Let us prove that

$$\lim_{\lambda \to \infty} F_\lambda f = f, \quad f \in \mathfrak{H}. \tag{6.30}$$

Since the family of the operators F_λ, $\lambda \in \mathbb{R}^1$, is monotone non-decreasing, the limit in the left-hand side of (6.30) exists. So it suffices to show that $F_\lambda \to I$ weakly, i.e.

$$\lim_{\lambda \to \infty} (F_\lambda f, g) = (f, g)$$

for any $f, g \in \mathfrak{H}$. Evidently, the last relation holds if

$$\lim_{\lambda \to \infty} (F_\lambda f, f) = (f, f), \quad f \in \mathfrak{H}. \tag{6.31}$$

But in view of (6.29) $\|F_\lambda\| \le 1$. Therefore we need only verify that (6.31) is fulfilled for a dense set of vectors from $\mathfrak{H}$. As this set we choose $\mathcal{D}(A)$. It is also rather obvious, by virtue of (6.28), that the equalities (6.31) and

$$\lim_{y \to \infty} iy \left(S_{iy} f, f \right) = -(f, f) \tag{6.32}$$

are equivalent. Thus, it remains to prove (6.32) for $f \in \mathcal{D}(A)$. Bearing in mind that the point $\left(S_{iy} f, f \right)$ belongs to the disk $K(f, iy)$, we get

$$\left| iy(S_{iy} f, f) + (f, f) \right| \le \max_{\tau \in \mathbb{R}^1} \left| iy(R_{iy}^\tau f, f) + (f, f) \right|,$$

where R_z^τ denotes the canonical resolvent associated with the parameter τ. Denote by A_τ the corresponding self-adjoint extension within $\mathfrak{H}$ of the operator A, and by E_λ^τ its spectral function. Then

$$\left| iy \left(R_{iy}^\tau f, f \right) + (f, f) \right|$$

$$= \left| \int_{-\infty}^{\infty} \frac{iy}{\lambda - iy} d(E_\lambda^\tau f, f) + \int_{-\infty}^{\infty} d(E_\lambda^\tau f, f) \right| = \left| \int_{-\infty}^{\infty} \frac{\lambda}{\lambda - iy} d(E_\lambda^\tau f, f) \right|$$

$$\le \left(\int_{-\infty}^{\infty} \lambda^2 \, d(E_\lambda^\tau f, f) \right)^{1/2} \left(\int_{-\infty}^{\infty} \frac{1}{|\lambda - iy|^2} d(E_\lambda^\tau f, f) \right)^{1/2} \le \frac{1}{y} \|Af\| \cdot \|f\|$$

if $f \in \mathcal{D}(A)$ and $y > 0$. Consequently,

$$iy \left(R_{iy}^\tau f, f \right) + (f, f) \to 0 \quad \text{as} \quad y \to \infty$$

uniformly with respect to $\tau \in \mathbb{R}^1$ as required. As we can see, F_λ is a resolution of the identity. By the Naimark theorem,

$$F_\lambda = P \tilde{E}_\lambda,$$

where $\widetilde{E}_\lambda$ is a certain orthogonal resolution of the identity in the space $\widetilde{\mathfrak{H}} \supset \mathfrak{H}$, P is the orthogonal projector from $\widetilde{\mathfrak{H}}$ onto $\mathfrak{H}$. Put

$$\widetilde{A}f = \int_{-\infty}^{\infty} \lambda \, d\widetilde{E}_\lambda f \,.$$

Obviously,

$$S_z = P\widetilde{R}_z \,,$$

where $\widetilde{R}_z$ is the resolvent of $\widetilde{A}$. To complete the proof we have to show that

$$\widetilde{A} \supset A$$

or, amounting to the same,

$$\widetilde{R}_z f = R_z f, \quad f \in \mathfrak{M}_z, \quad \Im z \neq 0 \,,$$

for an arbitrary generalized resolvent of the operator A. But it follows from (6.22) that

$$P\widetilde{R}_z f = S_z f = \overset{\circ}{R}_z f = R_z f$$

if $f \in \mathfrak{M}_z$. Hence,

$$\widetilde{R}_z f = R_z f + h, \quad h \in \widetilde{\mathfrak{H}} \ominus \mathfrak{H} \,.$$

However $h = 0$. Indeed, for $f \in \mathfrak{M}_z$

$$AR_z f = f + zR_z f \,,$$

whence

$$\widetilde{A}\big(R_z f + h\big) = f + zR_z f + zh = AR_z f + zh \,,$$

and we obtain

$$0 \leq \big(\widetilde{A}(R_z f + h), R_z f + h\big) = \big(AR_z f + zh, R_z f + h\big) = \big(AR_z f, R_z f\big) + z(h, h) \,,$$

contrary to the assumption $\Im z \neq 0$. Thus, $h = 0$. □

7   On Distribution Functions of a Hermitian Operator

7.1   Let $A\,(\overline{\mathcal{D}(A)} = \mathfrak{H})$ be a closed simple Hermitian operator whose defect numbers are equal to 1, and u the gauge of the representation (1.2.6). We recall that by a distribution function associated with the gauge u of the operator A we mean a function $\sigma(\lambda)$ of the form

$$\sigma(\lambda) = \big(E_\lambda u, u\big), \tag{7.1}$$

where E_λ is a spectral function of A. The class of all such functions is denoted by V_u. If a spectral function E_λ in (7.1) is orthogonal, then $\sigma(\lambda)$ is called canonical.

As Theorem 1.2.4 shows, the Stieltjes inversion formula determines a one-to-one correspondence between V_u and the set of all functions $w_u(z) = (R_z u\, u)$,

$$w(z) = w_u(z) = \int_{-\infty}^{\infty} \frac{d\sigma(\lambda)}{\lambda - z} = (R_z u\, u), \quad \sigma(\lambda) \in V_u, \tag{7.2}$$

(R_z is a generalized resolvent of the operator A).

On the other hand, if we take $f = u$ in the formula (6.20) and multiply the two sides of this equality scalarly by u, then after reducing to the common denominator and changing $-\frac{1}{\tau(z)}$ for $\tau(z)$ we get

$$w(z) = \frac{p_1(z)\tau(z) + p_0(z)}{q_1(z)\tau(z) + q_0(z)}, \tag{7.3}$$

where

$$q_0(z) = -\frac{1}{(u\,, \varphi(\bar z))}, \quad q_1(z) = -Q_1(z)q_0(z),$$

$$p_0(z) = \overset{\circ}{w}(z)q_0(z), \quad p_1(z) = \overset{\circ}{w}(z)q_1(z) + (\varphi(z), u),$$

$\overset{\circ}{w}(z) = (\overset{\circ}{R}_z u\,, u)$, $\overset{\circ}{R}_z$ is a fixed canonical resolvent of the operator A, $\tau(z)$ is an R-function.

Thus, the formulas (7.2) and (7.3) establish a one-to-one correspondence between the set of all R-functions $\tau(z)$ including $\tau(z) \equiv \infty$, and the set V_u of all distribution functions $\sigma(\lambda)$ of the operator A so that the functions $\tau(z) \equiv t$, $t = $ const, $t \in \mathbb{R}^1$, are associated with canonical ones.

It should be noted that the functions $q_0(z)$, $p_0(z)$, $q_1(z)$, $p_1(z)$ do not depend on the choice of a generalized resolvent R_z. They are analytic at every u-regular point of the operator A.

7.2 Let $\{d_k\}_{k=0}^{\infty}$ be an orthonormal basis in $\mathfrak{H}$ and $\mathcal{D}_k(z)u = d_{ku}(z)u$ the image of d_k in the isomorphism (1.2.6). According to (1.2.5)

$$\mathcal{D}_k(z) = \frac{(d_k, \varphi(\bar z))}{(u\,, \varphi(\bar z))}.$$

As

$$d_k - \mathcal{D}_k(z)u \in \mathfrak{M}_z,$$

we have

$$d_k - \mathcal{D}_k(z)u = (A - zI)e_k(z), \quad e_k(z) \in \mathcal{D}(A).$$

The vector-functions $e_k(z)$ are analytic at any u-regular point of the operator A. Indeed, if z is a u-regular point of A, then, by Theorem 3.1, z is a point of regular type for this operator. In view of Proposition 1.3.1, there exists a self-adjoint extension $\widetilde{A}$ within $\mathfrak{H}$ of the operator A such that the point z is regular for this extension. Then

$$e_k(z) = \widetilde{R}_z\big(d_k - \mathcal{D}_k(z)u\big)\,,$$

where $\widetilde{R}_z$ is the resolvent of the operator $\widetilde{A}$, whence the analyticity of the function $e_k(z)$ follows.

Put

$$\mathcal{E}_k(z) = \Big(R_z\big(d_k - \mathcal{D}_k(z)u\big), u\Big) = \int\limits_{-\infty}^{\infty} \frac{\mathcal{D}_k(\lambda) - \mathcal{D}_k(z)}{\lambda - z}\, d\sigma(\lambda)\,. \qquad (7.4)$$

It is not hard to see that $\mathcal{D}_k(z)$ and $\mathcal{E}_k(z)$ do not depend on the choice of a generalized resolvent R_z. They are analytic at every u-regular point of A.

7.3 It follows from (7.3) and Theorem 6.5 that for every fixed $z : \Im z > 0$ the points

$$w(z) = \int\limits_{-\infty}^{\infty} \frac{d\sigma(\lambda)}{\lambda - z}$$

completely fill the disk $K(z) = K(u\,,z)$ (the so-called Weyl disk) when $\sigma(\lambda)$ passes through the set V_u of all distribution functions of the operator A. The boundary $C(z) = C(u,\,z)$ of this disk is situated on the upper half-plane. The point $w = w(z)$ belongs to the circle $C(z)$ if and only if it corresponds to a canonical $\sigma(\lambda)$. When $\sigma(\lambda)$ goes through the set of all the canonical distribution functions of A, the points w corresponding to them fill the whole circle $C(z)$.

Let us write the equation of $C(z)$ in terms of the functions $\mathcal{D}_k(z)$ and $\mathcal{E}_k(z)$. By virtue of the Hilbert identity for canonical resolvents we have

$$\frac{w(z) - w(\overline{\zeta})}{z - \overline{\zeta}} = \frac{\big((R_z - R_{\overline{\zeta}})u,\, u\big)}{z - \overline{\zeta}} = (R_z R_{\overline{\zeta}} u\,,\, u) = (R_{\overline{\zeta}} u\,,\, R_{\overline{z}} u)$$

$$= \sum_{k=0}^{\infty} (R_{\overline{\zeta}} u\,,\, d_k)\big(d_k,\, R_{\overline{z}} u\big) = \sum_{k=0}^{\infty} (u\,,\, R_{\zeta} d_k)\big(R_z d_k,\, u\big)\,.$$

Since

$$(R_z d_k,\, u) = \Big(R_z\big(d_k - \mathcal{D}_k(z)u\big),\, u\Big) + w(z)\mathcal{D}_k(z) = \mathcal{E}_k(z) + w(z)\mathcal{D}_k(z)\,,$$

the equality

$$\frac{w(z) - w(\overline{\zeta})}{z - \overline{\zeta}} = \sum_{k=0}^{\infty} \overline{\big(\mathcal{E}_k(\zeta) + w(\zeta)\mathcal{D}_k(\zeta)\big)}\big(\mathcal{E}_k(z) + w(z)\mathcal{D}_k(z)\big) \qquad (7.5)$$

is valid. The substitution of z for ζ in (7.5) yields

$$\frac{w(z) - w(\bar{z})}{z - \bar{z}} = \sum_{k=0}^{\infty} \left| \mathcal{E}_k(z) + w(z) \mathcal{D}_k(z) \right|^2 . \tag{7.6}$$

The series on the right-hand side of the equation (7.6) converges uniformly in some neighbourhood of every u-regular point z. In fact, since $\mathcal{D}_k(z)$ and $\mathcal{E}_k(z)$ do not depend on the choice of a self-adjoint extension, in the case when z is a real u-regular point of A we may take, in view of Proposition 1.3.1, the extension $\widetilde{A}$ within $\mathfrak{H}$ such that z is regular for this extension. Then the function

$$\frac{w(z) - w(\bar{z})}{z - \bar{z}}$$

corresponding to $\widetilde{A}$ is continuous in a certain neighbourhood of the point z. By the well-known Dini theorem, the series (7.6) converges uniformly in this neighbourhood. If the point z is non-real, then the uniform convergence of the series (7.6) in some neighbourhood of z is obvious due the same Dini theorem. Remembering Theorem 4.1, we can now conclude that together with the series $\sum_{k=0}^{\infty} \left| \mathcal{D}_k(z) \right|^2$ the series $\sum_{k=0}^{\infty} \left| \mathcal{E}_k(z) \right|^2$ converges uniformly in a certain neighbourhood of z. Therefore the equation (7.6) may be written in the form

$$\alpha w \bar{w} + \beta \bar{w} + \bar{\beta} w + \gamma = 0 , \quad \text{or, equivalently,} \quad \alpha \left| w + \frac{\beta}{\alpha} \right|^2 = \frac{|\beta|^2}{\alpha} - \gamma ,$$

where

$$\alpha = \sum_{k=0}^{\infty} \left| \mathcal{D}_k(z) \right|^2, \quad \beta = \sum_{k=0}^{\infty} \overline{\mathcal{D}_k(z)} \mathcal{E}_k(z) + \frac{1}{z - \bar{z}} , \quad \gamma = \sum_{k=0}^{\infty} \left| \mathcal{E}_k(z) \right|^2 .$$

Thus, we have obtained the equation of the circle $C(z)$, whose coefficients are expressed in terms of $\mathcal{D}_k(z)$ and $\mathcal{E}_k(z)$. Its center and radius are determined by the formulas

$$\zeta(z) = \frac{i - 2y \sum_{k=0}^{\infty} \overline{\mathcal{D}_k(z)} \mathcal{E}_k(z)}{2y \sum_{k=0}^{\infty} \left| \mathcal{D}_k(z) \right|^2} ;$$

$$r(z) = \frac{1}{2y \left(\sum_{k=0}^{\infty} \left| \mathcal{D}_k(z) \right|^2 \sum_{k=0}^{\infty} \left| \mathcal{D}_k(\bar{z}) \right|^2 \right)^{1/2}} .$$

The disk $K(z)$ is given by the inequality

$$\frac{w(z) - w(\bar{z})}{z - \bar{z}} \geq \sum_{k=0}^{\infty} \left| \mathcal{E}_k(z) + w(z) \mathcal{D}_k(z) \right|^2 .$$

As the series $\sum\limits_{k=1}^{\infty} |\mathcal{D}_k(z)|^2$ and $\sum\limits_{k=1}^{\infty} |\mathcal{E}_k(z)|^2$ converge uniformly in some neighbourhoods of u-regular points z and ζ of the operator A, it is possible to reduce the formula (7.5) to the form

$$w(z) = \frac{\mathcal{P}_1(z,\zeta)w(\zeta) + \mathcal{P}_0(z,\zeta)}{\mathcal{Q}_1(z,\zeta)w(\zeta) + \mathcal{Q}_0(z,\zeta)} , \tag{7.7}$$

where the functions

$$\mathcal{Q}_1(z,\zeta) = -(z-\zeta) \sum_{k=0}^{\infty} \mathcal{D}_k(z)\overline{\mathcal{D}_k(\bar\zeta)} ,$$

$$\mathcal{Q}_0(z,\zeta) = 1 - (z-\zeta) \sum_{k=0}^{\infty} \mathcal{D}_k(z)\overline{\mathcal{E}_k(\bar\zeta)} ,$$

$$\mathcal{P}_1(z,\zeta) = 1 + (z-\zeta) \sum_{k=0}^{\infty} \mathcal{E}_k(z)\overline{\mathcal{D}_k(\bar\zeta)} ,$$

$$\mathcal{P}_0(z,\zeta) = (z-\zeta) \sum_{k=0}^{\infty} \mathcal{E}_k(z)\overline{\mathcal{E}_k(\bar\zeta)}$$

$$\tag{7.8}$$

are analytic in each variable at every u-regular point of the operator A.

7.4 Setting $z = \zeta$ and $\tau(z) \equiv t$ in the equation (7.3), where t is a real constant, and solving this equation with respect to t, we find due to the reality of the function

$$w(\zeta) = \int_{-\infty}^{\infty} \frac{d\sigma(\lambda)}{\lambda - \zeta} , \quad \text{that} \quad t = \frac{\overline{p_0(\bar\zeta)} - \overline{q_0(\bar\zeta)}\, w(\zeta)}{-\overline{p_1(\bar\zeta)} + \overline{q_1(\bar\zeta)}\, w(\zeta)} .$$

The substitution of this expression for $\tau(z) \equiv t$ in (7.3) gives

$$w(z) = \frac{a_1(z,\zeta)\, w(\zeta) + a_0(z,\zeta)}{b_1(z,\zeta)\, w(\zeta) + b_0(z,\zeta)} . \tag{7.9}$$

Here the matrix of the coefficients is determined by the relation

$$\begin{pmatrix} a_1(z,\zeta) & a_0(z,\zeta) \\ b_1(z,\zeta) & b_0(z,\zeta) \end{pmatrix} = \begin{pmatrix} p_1(z) & p_0(z) \\ q_1(z) & q_0(z) \end{pmatrix} \begin{pmatrix} -\overline{q_0(\bar\zeta)} & \overline{p_0(\bar\zeta)} \\ \overline{q_1(\bar\zeta)} & -\overline{p_1(\bar\zeta)} \end{pmatrix} .$$

Comparing (7.7) and (7.9), and taking into account that the point $w(\zeta)$ passes through the whole circle $C(\zeta)$ while $\sigma(\lambda)$ runs through the set of all canonical distribution functions of the operator A, we arrive at the equality

$$\begin{pmatrix} \mathcal{P}_1(z,\zeta) & \mathcal{P}_0(z,\zeta) \\ \mathcal{Q}_1(z,\zeta) & \mathcal{Q}_0(z,\zeta) \end{pmatrix} = c(z,\zeta) \begin{pmatrix} p_1(z) & p_0(z) \\ q_1(z) & q_0(z) \end{pmatrix} \begin{pmatrix} -\overline{q_0(\bar\zeta)} & \overline{p_0(\bar\zeta)} \\ \overline{q_1(\bar\zeta)} & -\overline{p_1(\bar\zeta)} \end{pmatrix} ,$$

or, equivalently,

$$
\begin{aligned}
\mathcal{P}_1(z,\zeta) &= c(z,\zeta)\big(-p_1(z)\overline{q_0(\overline{\zeta})} + p_0(z)\overline{q_1(\overline{\zeta})}\big)\,,\\
\mathcal{P}_0(z,\zeta) &= c(z,\zeta)\big(p_1(z)\overline{p_0(\overline{\zeta})} - p_0(z)\overline{p_1(\overline{\zeta})}\big)\,,\\
\mathcal{Q}_1(z,\zeta) &= c(z,\zeta)\big(-q_1(z)\overline{q_0(\overline{\zeta})} + q_0(z)\overline{q_1(\overline{\zeta})}\big)\,,\\
\mathcal{Q}_0(z,\zeta) &= c(z,\zeta)\big(q_1(z)\overline{p_0(\overline{\zeta})} - q_0(z)\overline{p_1(\overline{\zeta})}\big)\,.
\end{aligned}
\tag{7.10}
$$

In order to determine $c(z,\zeta)$ we first find the expression

$$
q_0(z)\overline{q_1(\overline{\zeta})} - q_1(z)\overline{q_0(\overline{\zeta})}\,.
$$

We have:

$$
\begin{aligned}
q_0(z)\overline{q_1(\overline{\zeta})} - q_1(z)\overline{q_0(\overline{\zeta})} &= q_0(z)\overline{q_0(\overline{\zeta})}\big(Q_1(z) - Q_1(\zeta)\big) =\\
&= q_0(z)\overline{q_0(\overline{\zeta})}(z-\zeta)\big(\varphi(\zeta),\varphi(\overline{z})\big)\\
&= (z-\zeta)\,q_0(z)\overline{q_0(\overline{\zeta})}\sum_{k=0}^{\infty}\big(\varphi(\zeta),d_k\big)\big(d_k,\varphi(\overline{z})\big)\\
&= (z-\zeta)\sum_{k=0}^{\infty}\frac{\big(d_k,\varphi(\overline{z})\big)}{\big(u,\varphi(\overline{z})\big)}\cdot\frac{\overline{\big(d_k,\varphi(\overline{\zeta})\big)}}{\overline{\big(u,\varphi(\overline{\zeta})\big)}}\\
&= (z-\zeta)\sum_{k=0}^{\infty}\mathcal{D}_k(z)\overline{\mathcal{D}_k(\overline{\zeta})} = -Q_1(z,\zeta)\,.
\end{aligned}
$$

The third equation in (7.10) gives the identity

$$
c(z,\zeta) \equiv -1\,.
$$

Now, in view of (7.8), the formulas (7.10) take the form

$$
q_0(z)\overline{q_1(\overline{\zeta})} - q_1(z)\overline{q_0(\overline{\zeta})} = (z-\zeta)\sum_{k=0}^{\infty}\mathcal{D}_k(z)\overline{\mathcal{D}_k(\overline{\zeta})}\,,
$$

$$
q_0(z)\overline{p_1(\overline{\zeta})} - q_1(z)\overline{p_0(\overline{\zeta})} = 1 - (z-\zeta)\sum_{k=0}^{\infty}\mathcal{D}_k(z)\overline{\mathcal{E}_k(\overline{\zeta})}\,,
$$

$$
p_0(z)\overline{q_1(\overline{\zeta})} - p_1(z)\overline{q_0(\overline{\zeta})} = -1 - (z-\zeta)\sum_{k=0}^{\infty}\mathcal{E}_k(z)\overline{\mathcal{D}_k(\overline{\zeta})}\,,
\tag{7.11}
$$

$$
p_0(z)\overline{p_1(\overline{\zeta})} - p_1(z)\overline{p_0(\overline{\zeta})} = (z-\zeta)\sum_{k=0}^{\infty}\mathcal{E}_k(z)\overline{\mathcal{E}_k(\overline{\zeta})}\,.
$$

These equalities are valid for arbitrary u-regular points z and ζ of the operator A. Moreover, they show that in spite of dependence of the functions $D_k(z)$ and $\mathcal{E}_k(z)$ on the choice of a basis $\{d_k\}_{k=0}^{\infty}$, the sums $\sum_{k=0}^{\infty} |D_k(z)|^2$ and $\sum_{k=0}^{\infty} |\mathcal{E}_k(z)|^2$ are independent of this choice.

7.5   Since the function $w(z)$ is real, the relation

$$\frac{p_1(z)\, t + p_0(z)}{q_1(z)\, t + q_0(z)} = \frac{\overline{p_1(\overline{z})}\, t + \overline{p_0(\overline{z})}}{\overline{q_1(\overline{z})}\, t + \overline{q_0(\overline{z})}}$$

holds, where t is any real constant. Therefore

$$\frac{p_1(z)}{\overline{p_1(\overline{z})}} = \frac{p_0(z)}{\overline{p_0(\overline{z})}} = \frac{q_1(z)}{\overline{q_1(\overline{z})}} = \frac{q_0(z)}{\overline{q_0(\overline{z})}}. \tag{7.12}$$

Let a be a real u-regular point of A. For the sake of simplicity we may assume $a = 0$. The substitution of $z = a = 0$ in (7.12) yields

$$\frac{p_1(0)}{\overline{p_1(0)}} = \frac{p_0(0)}{\overline{p_0(0)}} = \frac{q_1(0)}{\overline{q_1(0)}} = \frac{q_0(0)}{\overline{q_0(0)}} = \varepsilon, \quad |\varepsilon| = 1.$$

Dividing the functions $p_1(z)$, $p_0(z)$, $q_1(z)$, $q_0(z)$ by $\sqrt{\varepsilon}$, it is possible to make real the values $p_1(0)$, $p_0(0)$, $q_1(0)$, $q_0(0)$. In so doing we do not violate the identities (7.11). Setting $z = \zeta$ in the third equation in (7.11), we obtain

$$p_0(z)\overline{q_1(\overline{z})} - p_1(z)\overline{q_0(\overline{z})} = -1.$$

It follows that the matrix

$$\begin{pmatrix} p_1(0) & p_0(0) \\ q_1(0) & q_0(0) \end{pmatrix}$$

is real, and its determinant equals 1.

Suppose

$$t = \frac{\alpha\, t' + \beta}{\gamma\, t' + \delta},$$

where the real numbers α, β, γ, δ satisfy the condition $\alpha\delta - \beta\gamma = 1$. Then

$$w(z) = \frac{p_1'(z)\, t' + p_0'(z)}{q_1'(z)\, t' + q_0'(z)}.$$

The coefficients $p_1'(z)$, $p_0'(z)$, $q_1'(z)$ and $q_0'(z)$ are related to $p_1(z)$, $p_0(z)$, $q_1(z)$, $q_0(z)$ by the formula

$$\begin{pmatrix} p_1'(z) & p_0'(z) \\ q_1'(z) & q_0'(z) \end{pmatrix} = \begin{pmatrix} p_1(z) & p_0(z) \\ q_1(z) & q_0(z) \end{pmatrix} \begin{pmatrix} \alpha & \beta \\ \gamma & \delta \end{pmatrix}.$$

In particular,

$$q_1'(z) = \alpha\, q_1(z) + \gamma\, q_0(z)\,,$$
$$q_0'(z) = \beta\, q_1(z) + \delta\, q_0(z)\,.$$

The direct computation gives

$$q_0'(z)\overline{q_1'(\zeta)} - q_1'(z)\overline{q_0'(\zeta)} = q_0(z)\overline{q_1(\zeta)} - q_1(z)\overline{q_0(\zeta)}\,,$$

which shows that the formulas (7.11) are also true for the functions $p_1'(z)$, $p_0'(z)$, $q_1'(z)$, $q_0'(z)$. In addition

$$\det\begin{pmatrix} p_1'(z) & p_0'(z) \\ q_1'(z) & q_0'(z) \end{pmatrix} = \det\begin{pmatrix} p_1(z) & p_0(z) \\ q_1(z) & q_0(z) \end{pmatrix}\,,$$

whence

$$\det\begin{pmatrix} p_1'(0) & p_0'(0) \\ q_1'(0) & q_0'(0) \end{pmatrix} = 1\,.$$

Choose α, β, γ, δ so that

$$\begin{pmatrix} p_1'(0) & p_0'(0) \\ q_1'(0) & q_0'(0) \end{pmatrix} = \begin{pmatrix} 1 & 0 \\ 0 & 1 \end{pmatrix}\,. \tag{7.13}$$

Since the map

$$\tau(z) = \frac{\alpha\,\tau'(z) + \beta}{\gamma\,\tau'(z) + \delta}$$

transforms the class of all R-functions onto itself, we may assume from the very beginning that the functions $p_0(z)$, $p_1(z)$, $q_0(z)$, $q_1(z)$ in (7.3) possess the property (7.13). Putting $\zeta = 0$ in (7.11), we arrive at the following effective formulas for $p_0(z)$, $p_1(z)$, $q_0(z)$, $q_1(z)$:

$$q_1(z) = -z \sum_{k=1}^{\infty} \mathcal{D}_k(z)\overline{\mathcal{D}_k(0)}\,,$$

$$q_0(z) = 1 - z \sum_{k=1}^{\infty} \mathcal{D}_k(z)\overline{\mathcal{E}_k(0)}\,,$$

$$p_1(z) = 1 + z \sum_{k=1}^{\infty} \mathcal{E}_k(z)\overline{\mathcal{D}_k(0)}\,, \tag{7.14}$$

$$p_0(z) = z \sum_{k=1}^{\infty} \mathcal{E}_k(z)\overline{\mathcal{E}_k(0)}\,.$$

7.6 Let J be an operator given on the whole space $\mathfrak{H}$.

Definition 7.1 *The operator J is called an involution on $\mathfrak{H}$ if*

$$J^2 = I, \quad J(\alpha f + \beta g) = \overline{\alpha}Jf + \overline{\beta}Jg \quad and \quad (Jf, Jg) = \overline{(f,g)}$$

for any vectors $f, g \in \mathfrak{H}$, and any complex numbers α, β.

Definition 7.2 *We call a Hermitian operator A on $\mathfrak{H}$ real with respect to the involution J if*

$$J\mathcal{D}(A) \subseteq \mathcal{D}(A) \quad and \quad JAf = AJf, \quad f \in \mathcal{D}(A).$$

An operator A real with respect to the involution J possesses the following properties:

(i) $J(A - zI) = (A - \overline{z}I)J$;

(ii) $J\mathfrak{M}_z = \mathfrak{M}_{\overline{z}}$;

(iii) $JR_z(A) = R_{\overline{z}}(A)J$;

here $R_z(A)$ is the resolvent of the operator A.

Lemma 7.1 *Let $A\left(\overline{\mathcal{D}(A)} = \mathfrak{H}\right)$ be a closed simple Hermitian operator whose defect index is (1,1), and $\overset{\circ}{A}$ its self-adjoint extension within $\mathfrak{H}$. There exists an involution on $\mathfrak{H}$ such that $\overset{\circ}{A}$ is real with respect to this involution.*

Proof. Having fixed a self-adjoint extension $\overset{\circ}{A}$ within $\mathfrak{H}$ of the operator A, construct the family $\varphi(z)$ as was done in (1.2.3). We define the operator J on the elements $\varphi(z)$ as

$$J\varphi(z) = \varphi(\overline{z}),$$

and then extend it to linear combinations of $\varphi(z)$ in the following way:

$$J\sum_{k=1}^{n} c_k\varphi(z_k) = \sum_{k=1}^{n} \overline{c}_k\varphi(\overline{z}_k).$$

Since

$$\left(J\varphi(z), J\varphi(\zeta)\right) = \left(\varphi(\overline{z}), \varphi(\overline{\zeta})\right)$$
$$= \frac{Q_1(\zeta) - Q_1(\overline{z})}{\zeta - \overline{z}} = \frac{\overline{Q_1(\overline{\zeta})} - \overline{Q_1(z)}}{\overline{\zeta} - z} = \overline{\left(\varphi(z), \varphi(\zeta)\right)},$$

where $Q_1(z)$ is defined by (6.19), and the linear span of $\varphi(z)$ is dense in $\mathfrak{H}$ (the operator A is simple), the closure of J is an involution on $\mathfrak{H}$. Taking into account that

$$J\overset{\circ}{R}_z\varphi(z) = J\frac{\varphi(z) - \varphi(\zeta)}{z - \zeta} = \frac{\varphi(\overline{z}) - \varphi(\overline{\zeta})}{z - \zeta} = \overset{\circ}{R}_{\overline{z}}\varphi(\overline{\zeta}) = \overset{\circ}{R}_{\overline{z}}J\varphi(\zeta),$$

we get, by passing to the limit,

$$J \overset{\circ}{R_z} f = \overset{\circ}{R_{\bar{z}}} J f \tag{7.15}$$

for any $f \in \mathfrak{H}$. The equality (7.15) is equivalent to

$$J \overset{\circ}{A} f = \overset{\circ}{A} J f, \quad f \in \mathfrak{H},$$

which is what had to be proved. $\qquad\square$

Theorem 7.1 *For every closed simple Hermitian operator A on $\mathfrak{H}$ $(\overline{\mathcal{D}(A)} = \mathfrak{H})$ whose deficiency index is $(1,1)$, there exists an involution J on $\mathfrak{H}$ such that the operator A is real with respect to this involution.*

Proof. The theorem is a direct consequence of Lemma 7.1. In fact, suppose $f \in \mathfrak{M}_z$, that is $(f, \varphi(\bar{z})) = 0$. Then

$$\big(Jf, \varphi(z)\big) = \overline{\big(J^2 f, J\varphi(z)\big)} = \overline{\big(f, \varphi(\bar{z})\big)} = 0 \,.$$

Hence,

$$J\mathfrak{M}_z \subseteq \mathfrak{M}_{\bar{z}} \,. \tag{7.16}$$

In view of (7.15) and (7.16),

$$JR_z(A)f = J\overset{\circ}{R_z}f = \overset{\circ}{R_{\bar{z}}}Jf \in \mathcal{D}(A)$$

for $f \in \mathfrak{M}_z$, whence $J\mathcal{D}(A) \subseteq \mathcal{D}(A)$. Besides,

$$JAg = J \overset{\circ}{A} g = \overset{\circ}{A} Jg = AJg$$

if $g \in \mathcal{D}(A)$, which completes the proof. $\qquad\square$

Definition 7.3 *A vector $f \in \mathfrak{H}$ is called real with respect to an involution J if $Jf = f$.*

In a Hilbert space with involution J it is always possible to choose an orthonormal basis $\{d_k\}_{k=0}^{\infty}$ real in this involution. The example is the basis obtained by the standard orthogonalization of the sequence $\left\{ \dfrac{e_k + Je_k}{2}, \dfrac{e_k - Je_k}{2i} \right\}_{k \in \mathbb{N}_0}$, where $\{e_k\}_{k \in \mathbb{N}}$ is an arbitrary orthonormal basis in $\mathfrak{H}$.

Let J be the involution introduced in Theorem 7.1. Take the gauge u real with respect to J. We can always do this, setting, for instance, $u = \varphi(i) + \varphi(-i) = Ju$. The existence of $z : \Im z \neq 0$ such that $(u, \varphi(z)) \neq 0$ is easily seen in view of the linear independence of the vectors $\varphi(i)$ and $\varphi(-i)$ and the simplicity of A. Since

$$\mathcal{D}_k(\bar{z}) = \frac{(d_k, \varphi(z))}{(u, \varphi(z))} = \frac{\overline{(Jd_k, J\varphi(z))}}{\overline{(Ju, J\varphi(z))}} = \frac{(d_k, \varphi(\bar{z}))}{(u, \varphi(\bar{z}))} = \overline{\mathcal{D}_k(z)} \,,$$

the functions $\mathcal{D}_k(z)$, $k \in \mathbb{N}_0$, are real, and so are the functions $\mathcal{E}_k(z)$, $k \in \mathbb{N}$. Indeed,

$$\mathcal{E}(\bar{z}) = \left(R_{\bar{z}}(d_k - \mathcal{D}_k(\bar{z}))u\,,u\right) = \overline{\left(JR_{\bar{z}}(d_k - \mathcal{D}_k(\bar{z}))u\,,u\right)}$$

$$= \overline{\left(R_z J(d_k - \mathcal{D}_k(\bar{z}))u\,,u\right)} = \left(R_z(d_k - \mathcal{D}_k(z))u\,,u\right) = \mathcal{E}_k(z)\,.$$

Therefore the functions $q_0(z)$, $q_1(z)$, $p_0(z)$, $p_1(z)$ appearing in (7.3), which are expressed in terms of $\mathcal{D}_k(z)$ and $\mathcal{E}_k(z)$ by the formula (7.14), are real too.

7.7 Suppose A to be an entire J-real operator with respect to a J-real entire gauge u. Such a gauge always exists. Then the functions $q_0(z)$, $q_1(z)$, $p_0(z)$, $p_1(z)$, $\mathcal{D}_k(z)$, $\mathcal{E}_k(z)$ are real entire ones, and the series $\sum\limits_{k=0}^{\infty} |\mathcal{D}_k(z)|^2$ and $\sum\limits_{k=0}^{\infty} |\mathcal{D}_k(z)|^2$ converge uniformly in every bounded domain of the complex plane.

It should be noted that the type of each of the functions $q_0(z)$, $q_1(z)$, $p_0(z)$, $p_1(z)$ is at most exponential. In fact, the function $(u\,,\varphi(\bar{z}))$ is entire and vanishes nowhere. By virtue of (2.2) and Lemma 1.4.4, it is an N-function in the upper and lower half-planes. Since the class of all N-functions forms a ring, $q_0(z)$ is an N-function there. By Theorem 1.4.5 it is of at most exponential type. The representation (6.19) and Lemma 1.4.4 show that the function $Q_1(z)$ is an N-function in both planes $\Im z > 0$ and $\Im z < 0$. Then the function

$$q_1(z) = -Q_1(z)q_0(z)$$

is also an N-function in these half-planes. Theorem 1.4.5 shows that the type of the function $q_1(z)$ is exponential. Similar arguments are applicable to the functions

$$p_0(z) = \left(\overset{\circ}{R}_z u\,,u\right)q_0(z) \quad \text{and} \quad p_1(z) = \left(\overset{\circ}{R}_z u\,,u\right)q_1(z) + (\varphi(z),u)\,.$$

Set

$$U(z) = \begin{pmatrix} p_1(z) & p_0(z) \\ q_1(z) & q_0(z) \end{pmatrix}\,.$$

In the case under consideration the matrix $U(z)$ possesses the properties:

1) the entries of $U(z)$ are real entire functions;

2) $U(0) = \begin{pmatrix} 1 & 0 \\ 0 & 1 \end{pmatrix}$;

3) $\det U(z) \equiv 1$;

4) for any fixed $z : \Im z > 0$ the linear-fractional function

$$\frac{p_1(z)\,t + p_0(z)}{q_1(z)\,t + q_0(z)}\,, \quad t \in \mathbb{R}^1\,,$$

maps the real axis into a certain circle lying in the upper half-plane.

Definition 7.4 *A matrix $U(z)$ satisfying the conditions 1)–4) is called a Nevanlinna matrix.*

The preceding reasoning leads to the following fundamental theorem.

Theorem 7.2 *Let A be an entire operator on $\mathfrak{H}$ whose deficiency index is $(1,1)$. Then a Nevanlinna matrix*

$$U(z) = \begin{pmatrix} p_1(z) & p_0(z) \\ q_1(z) & q_0(z) \end{pmatrix},$$

whose entries $p_1(z)$, $p_0(z)$, $q_1(z)$, $q_0(z)$ are entire functions of at most exponential type, corresponds to this operator so that the formula

$$\int_{-\infty}^{\infty} \frac{d\sigma(\lambda)}{\lambda - z} = \frac{p_1(z)\,\tau(z) + p_0(z)}{q_1(z)\,\tau(z) + q_0(z)} \qquad (7.17)$$

determines a one-to-one correspondence between the class V_u of all distribution functions $\sigma(\lambda)$ of the operator A and the set of all R-functions $\tau(z)$. The function $\sigma(\lambda)$ in (7.17) is canonical if and only if $\tau(z) \equiv t$, where t is a real constant, including $\tau(z) \equiv \infty$.

It is seen from (7.17) that if

$$U(z) = \begin{pmatrix} p_1(z) & p_0(z) \\ q_1(z) & q_0(z) \end{pmatrix}$$

is a Nevanlinna matrix, then the asymptotic relation

$$\frac{p_1(z)\,t + p_0(z)}{q_1(z)\,t + q_0(z)} = -\frac{s_0}{z} + o\!\left(\frac{1}{z}\right), \qquad z = iy, \quad y \in \mathbb{R}^1,$$

is valid, where

$$s_0 = \int_{-\infty}^{\infty} d\sigma(\lambda),$$

and $|y|$ is sufficiently large. Setting $t = \infty$, we conclude that

$$\frac{p_1(z)}{q_1(z)} = -\frac{s_0}{z} + o\!\left(\frac{1}{z}\right).$$

7.8 Let us turn our attention to the properties of the matrix $U(z)$. To study them we need the next assertion.

Lemma 7.2 *In order that the function*

$$w(z) = \frac{\alpha z + \beta}{\gamma z + \delta}, \qquad \alpha\delta - \gamma\beta = 1, \tag{7.18}$$

map the half-plane $\Im z \geq 0$ onto a disk lying completely on the half-plane $\Im w > 0$, it is necessary and sufficient that the linear transform

$$\begin{aligned} w_1 &= \alpha z_1 + \beta z_2 \\ w_2 &= \gamma z_1 + \delta z_2 \end{aligned} \tag{7.19}$$

possess the property

$$\frac{w_1\overline{w}_2 - w_2\overline{w}_1}{i} > \frac{z_1\overline{z}_2 - z_2\overline{z}_1}{i} \tag{7.20}$$

for arbitrary z_1 and z_2 not simultaneously equal to zero.

Proof. The sufficiency is easily verified. Indeed, if we put $z = \frac{z_1}{z_2}$, then

$$w = \frac{\alpha \frac{z_1}{z_2} + \beta}{\gamma \frac{z_1}{z_2} + \delta} = \frac{\alpha z_1 + \beta z_2}{\gamma z_1 + \delta z_2} = \frac{w_1}{w_2}.$$

In view of (7.20),

$$\Im w = \frac{w_1\overline{w}_2 - w_2\overline{w}_1}{2i|w_2|^2} > \frac{z_1\overline{z}_2 - z_2\overline{z}_1}{2i|w_2|^2} = \frac{\Im z}{|w_2|^2|z_2|^2} \geq 0.$$

Thus, the function $w = w(z)$ maps the half-plane $\Im z \geq 0$ into itself. If the map $z \mapsto w$ transformed the half-plane $\Im z \geq 0$ onto itself, the numbers α, β, γ, δ would be real (up to the same factor, I.I. Privalov[3]), hence

$$w_1\overline{w}_2 - w_2\overline{w}_1 = z_1\overline{z}_2 - z_2\overline{z}_1,$$

contrary to (7.20). The parallel translation $w = z + \beta$, $\beta > 0$ is also impossible because the condition (7.20) is not fulfilled in this case.

Proceed to the necessity. Denote by C the circle which is the image of the real axis $z = x$ under the map (7.18). At the lowest point $N : w_N = w(x_N)$ of C (whose ordinate is the smallest)

$$\frac{dw(x)}{dx}\bigg|_{x=x_N} = \frac{\partial u(x,0)}{\partial x}\bigg|_{x=x_N} = \frac{1}{(\gamma x_N + \delta)^2}$$

where $w(z) = w(x + iy) = u(x,y) + iv(x,y)$. Since $u(x,0)$ increases monotonically in some neighbourhood of the point x_N, we have

$$\frac{1}{(\gamma x_N + \delta)^2} > 0.$$

Consequently, the number $\gamma x_N + \delta$ is real, whence

$$x_N = -\frac{\delta - \bar{\delta}}{\gamma - \bar{\gamma}} \,.$$

Substituting x_N for z in (7.18), we find the point N:

$$w_N = \frac{1 + \beta\bar{\gamma} - \alpha\bar{\delta}}{\delta\bar{\gamma} - \gamma\bar{\delta}} \,.$$

Since $w(z)$ is finite as $\Im z \geq 0$, it is possible for $\gamma z + \delta$ to vanish only at $z : \Im z < 0$. It follows that the point $z = -\frac{\delta}{\gamma}$ belongs to the lower half-plane, i.e.

$$\Im\left(-\frac{\delta}{\gamma}\right) = \frac{1}{2i}\frac{\gamma\bar{\delta} - \delta\bar{\gamma}}{|\gamma|^2} < 0 \,. \tag{7.21}$$

This implies

$$\frac{1}{i}(\delta\bar{\gamma} - \gamma\bar{\delta}) > 0 \,.$$

On the other hand,

$$\Re\left(\alpha\bar{\delta} - \beta\bar{\gamma}\right) - 1 > 0 \tag{7.22}$$

because $\Im w_N > 0$.

Now consider the Hermitian form

$$\frac{w_1\bar{w}_2 - w_2\bar{w}_1}{i} - \frac{z_1\bar{z}_2 - z_2\bar{z}_1}{i}$$

$$= \frac{\beta\bar{\delta} - \delta\bar{\beta}}{i}|z_2|^2 + \frac{\beta\bar{\gamma} - \delta\bar{\alpha} + 1}{i}z_2\bar{z}_1 + \frac{\alpha\bar{\delta} - \gamma\bar{\beta} - 1}{i}z_1\bar{z}_2 + \frac{\alpha\bar{\gamma} - \gamma\bar{\alpha}}{i}|z_1|^2 \,. \tag{7.23}$$

Since $z = 0$ corresponds to the point $w = \frac{\beta}{\delta}$ of the upper falf-plane, the coefficient in the term with $|z_2|^2$ is positive. Besides,

$$\Delta = \begin{vmatrix} \beta\bar{\delta} - \delta\bar{\beta} & -\delta\bar{\alpha} + \beta\bar{\gamma} + 1 \\ \alpha\bar{\delta} - \gamma\bar{\beta} - 1 & \alpha\bar{\gamma} - \gamma\bar{\alpha} \end{vmatrix} = 2\left(\Re\left(\alpha\bar{\delta} - \beta\bar{\gamma}\right) - 1\right) \tag{7.24}$$

is also positive. Consequently, the expression in the left-hand side of (7.23) is positive. $\qquad\square$

Lemma 7.3 *If the function (7.18) maps the half-plane $\Im z \geq 0$ onto a disk lying completely in the half-plane $\Im z > 0$, then so do the functions obtained from (7.18) by the transposition of α and δ or β and γ.*

Proof. Suppose that the function $w(z)$ possesses the above conditions. We consider only the function

$$w'(z) = \frac{\delta z + \beta}{\gamma z + \alpha}.$$

The relations in (7.19) are transformed into

$$w_1' = \delta z_1 + \beta z_2$$
$$w_2' = \gamma z_1 + \alpha z_2.$$

Setting up the corresponding Hermitian form

$$\frac{w_1'\overline{w_2'} - w_2'\overline{w_1'}}{i} - \frac{z_1\overline{z_2} - z_2\overline{z_1}}{i}$$

$$= \frac{\beta\overline{\alpha} - \alpha\overline{\beta}}{i}|z_2|^2 + \frac{\beta\overline{\gamma} - \alpha\overline{\delta} + 1}{i}z_2\overline{z_1} + \frac{\delta\overline{\alpha} - \gamma\overline{\beta} - 1}{i}z_1\overline{z_2} + \frac{\delta\overline{\gamma} - \gamma\overline{\delta}}{i}|z_1|^2,$$

we observe that, by virtue of (7.21), (7.22) and (7.24), $\frac{1}{i}(\delta\overline{\gamma} - \gamma\overline{\delta})$ is positive and

$$\Delta' = 2(\Re(\alpha\overline{\delta} - \beta\overline{\gamma}) - 1) = \Delta > 0.$$

So this form is positive. □

7.9 Put

$$J = \begin{pmatrix} 0 & -1 \\ 1 & 0 \end{pmatrix}, \quad J^2 = -I.$$

Then the inequality (7.20) may be written in the form

$$\frac{1}{i}w^*Jw > \frac{1}{i}z^*Jz \quad \text{or} \quad iw^*Jw < iz^*Jz,$$

which is the same, where

$$w = \begin{pmatrix} w_1 \\ w_2 \end{pmatrix}, \quad z = \begin{pmatrix} z_1 \\ z_2 \end{pmatrix},$$

$*$ designates the passage to the adjoint. If we define the matrix U as

$$U = \begin{pmatrix} \alpha & \beta \\ \gamma & \delta \end{pmatrix},$$

then the last inequality is equivalent to

$$iU^*JU < iJ.$$

Hence, the matrix U is $i\mathcal{J}$-contractive. Moreover, it is symplectic, that is

$$U'\mathcal{J}U = \mathcal{J},$$

where U' is the matrix transposed to U. Indeed,

$$i\,U'\mathcal{J}U = i\begin{pmatrix} \alpha & \gamma \\ \beta & \delta \end{pmatrix}\begin{pmatrix} 0 & -1 \\ 1 & 0 \end{pmatrix}\begin{pmatrix} \alpha & \beta \\ \gamma & \delta \end{pmatrix} = i\mathcal{J}.$$

Thus, if the function

$$w(z) = \frac{\alpha z + \beta}{\gamma z + \delta}, \quad \alpha\delta - \gamma\beta = 1,$$

transforms the half-plane $\Im z \geq 0$ into a disk lying on the upper half-plane, then the matrix

$$U = \begin{pmatrix} \alpha & \beta \\ \gamma & \delta \end{pmatrix}$$

corresponding to $w(z)$ is $i\mathcal{J}$-contractive and symplectic.

7.10 Now turn to the real matrix-function

$$U(z) = \begin{pmatrix} p_1(z) & p_0(z) \\ q_1(z) & q_0(z) \end{pmatrix}$$

introduced in subsection 7.7. In view of the properties 1)–4) of this function, at every fixed $\zeta : \Im\zeta > 0$ the linear-fractional function

$$\frac{p_1(\zeta)\,z + p_0(\zeta)}{q_1(\zeta)\,z + q_0(\zeta)}$$

transforms the half-plane $\Im z > 0$ onto the interior of the circle

$$w(\zeta) = \frac{p_1(\zeta)\,t + p_0(\zeta)}{q_1(\zeta)\,t + q_0(\zeta)}$$

lying completely on the upper half-plane $\Im w > 0$. By the Potapov theorem (see, for instance, Gohberg and Krein [1]), $U(z)$ is the monodromy matrix of a certain canonical system

$$\frac{dU(t,z)}{dt} = z\mathcal{J}H(t)U(t,z), \quad U(0,z) = I, \quad 0 \leq t \leq l, \tag{7.25}$$

where the real symmetric matrix

$$H(t) = \begin{pmatrix} a(t) & b(t) \\ b(t) & c(t) \end{pmatrix}$$

is such that

(i) $a(t) \geq 0$, $c(t) \geq 0$;

(ii) $\det H(t) \geq 0$;

(iii) $a(t) + c(t) = 1$.

Thus, $U(z)$ is the value of the matriciant $U(t, z)$ of the Hamilton system (7.25) at the point $l : U(z) = U(l, z)$. The matrix $H(t)$ is uniquely determined if it obeys the normalization condition $\operatorname{Sp} H(t) = 1$, $0 \leq t \leq l$. The type of the function $U(z)$ is equal to

$$\int_0^l \sqrt{\det H(t)} \, dt \, .$$

7.11 If A is an entire operator whose deficiency index is (1,1), then the function

$$\nabla_M^{-1}(z) = \min_{f \in \mathfrak{H}} \frac{\|f\|}{\|f_M(z)\|}$$

is defined at all points $z \in \mathbb{C}^1$, and

$$0 < \nabla_M^{-1}(z) \leq 1$$

(see subsection 4.2). The following assertion holds.

Theorem 7.3 *The equality*

$$\nabla_M^{-2}(\xi) = \max_{\sigma \in V_u} \left(\sigma(\xi + 0) - \sigma(\xi - 0) \right)$$

is valid for any real number ξ.

Proof. By virtue of Theorem 1.1

$$\nabla_M^{-2}(\xi) = \min_{f \in \mathfrak{H}} \frac{\|f\|^2}{\|f_M(z)\|^2} = \min_{f \in \mathfrak{H}} \frac{\int_{-\infty}^{\infty} |f_u(\lambda)|^2 \, d\sigma(\lambda)}{|f_u(\xi)|^2}$$

$$\geq \min_{f \in \mathfrak{H}} \frac{|f_u(\xi)|^2 \left(\sigma(\xi + 0) - \sigma(\xi - 0) \right)}{|f_u(\xi)|^2} = \sigma(\xi + 0) - \sigma(\xi - 0) \, .$$

So,

$$\nabla_M^{-2}(\xi) \geq \sigma(\xi + 0) - \sigma(\xi - 0)$$

for an arbitrary function $\sigma(\lambda) \in V_u$.

Let $\widetilde{A}_\xi$, $\xi \in \mathbb{R}^1$, be the self-adjoint extension within $\mathfrak{H}$ of the operator A for which the number ξ is an eigenvalue (Proposition 1.3.6). The distribution function

$\sigma_\xi(\lambda)$ of A corresponding to $\widetilde{A}_\xi$ is a step-function. One of its jumps is realized at the point ξ. Setting $\sigma(\lambda) = \sigma_\xi(\lambda)$ in the formula (7.17), we automatically obtain the following relation for the function $\tau(z)$ associated with the extension $\widetilde{A}_\xi$:

$$\tau(z) \equiv t_\xi = -\frac{q_0(\xi)}{q_1(\xi)}.$$

The substitution of t_ξ and $\sigma_\xi(\lambda)$ for $\tau(z)$ and $\sigma(\lambda)$ in (7.17) gives the equivalence of two meromorphic functions having a prime pole at the point $z = \xi$. Equalizing the principal parts of these functions, we obtain the desired result with regard to the equalities (7.11). Namely,

$$\nabla_M^{-2}(\xi) = \frac{1}{\sum_{k=1}^{\infty} |\mathcal{D}_k(\xi)|^2} = \sigma(\xi+0) - \sigma(\xi-0). \qquad \square$$

8   Method of Directing Functionals

8.1   Let $\mathfrak{L}$ be an arbitrary linear set and A_0 a linear operator on $\mathfrak{L}$ whose domain $\mathcal{D}(A_0)$ is in general a proper part of $\mathfrak{L}: \mathcal{D}(A_0) \neq \mathfrak{L}$. Further, let $\Phi(f, \lambda)$, $\lambda \in \mathbb{R}^1$, be a one-parameter family of linear functionals on $\mathfrak{L}$.

Definition 8.1   *We call the family $\Phi(f, \lambda)$ a directing family of functionals (a directing functional in short) for the operator A_0 if it possesses the following properties:*

1) *for a fixed $f \in \mathfrak{L}$ the function $\Phi(f, \lambda)$ is analytic on the whole real axis;*

2) *there exists at least one $\lambda \in \mathbb{R}^1$ such that the functional*

$$\Phi(\cdot, \lambda) \neq 0 ;$$

3) *whatever $f_0 \in \mathfrak{L}$ and $\lambda_0 \in \mathbb{R}^1$, the equation*

$$A_0 f - \lambda_0 f = f_0 \qquad (8.1)$$

is solvable if and only if

$$\Phi(f_0, \lambda_0) = 0 . \qquad (8.2)$$

Let us give the example of an operator which has a directing functional.

Take the space of functions continuous on the interval $[0, a)$, $a \leq \infty$, with compact supports of the form $[\alpha, \beta]$, $\alpha \geq 0$, $\beta < a$, as $\mathfrak{L}$, and define the operator A_0 as follows:

$$(A_0 f)(x) = -\frac{df(x)}{dx}, \qquad \mathcal{D}(A_0) = \{f \in C^1([0, a)) : f(0) = 0\},$$

where $C^1\big([0,a)\big)$ denotes the set of all continuously differentiable functions on $[0,a)$ from $\mathfrak{L}$. In this case the equation (8.1) becomes

$$-\frac{df(x)}{dx} - \lambda_0 f(x) = f_0(x). \tag{8.3}$$

It is not difficult to verify that the solution of the equation (8.3) which satisfies the condition $f(0) = 0$ is given by the formula

$$f(x) = e^{-\lambda_0 x} \int_0^x e^{\lambda_0 t} f_0(t)\, dt\,,$$

and if $f_0(x) \equiv 0$ on $(a - \varepsilon, a)$, then

$$f(x) = e^{-\lambda_0 x} \int_0^a e^{\lambda_0 t} f_0(t)\, dt$$

at every point $x \in (a - \varepsilon, a)$. Taking into account that the solution $f(x)$ must vanish in the neighbourhood of the point a, we conclude that the condition

$$\int_0^a e^{\lambda_0 t} f_0(t)\, dt = 0$$

is necessary and sufficient for $f(x)$ to belong to $\mathfrak{L}$. Thus, the family

$$\Phi(f, \lambda) = \int_0^a e^{\lambda x} f(x)\, dx$$

is a directing functional for the operator A_0.

Similarly, it can be shown that the family of the functionals

$$\Phi(f, \lambda) = \int_{-\infty}^{\infty} e^{i\lambda x} f(x)\, dx$$

is directing for the operator A_0 generated by the differential expression $i\frac{d}{dx}$ in the space $\mathfrak{L}$ of functions $\varphi(x)$ with compact supports continuous on $\mathbb{R}^1$ on the set $\mathcal{D}(A_0) = C_0^1(\mathbb{R}^1)$ of all continuously differentiable functions from $\mathfrak{L}$.

The following properties of a directing functional of the operator A_0 are a direct consequence of Definition 8.1.

a) If $f \in \mathcal{D}(A_0)$, then

$$\Phi(A_0 f, \lambda) \equiv \lambda \Phi(f, \lambda), \quad \lambda \in \mathbb{R}^1\,.$$

Indeed, setting $g = A_0 f - \lambda f$, we obtain $\Phi(g, \lambda) = 0$, whence $\Phi(A_0 f, \lambda) = \lambda \Phi(f, \lambda)$.

b) If

$$\Phi(f, \lambda_0) = 0 \tag{8.4}$$

for some $f \in \mathfrak{L}$ and $\lambda_0 \in \mathbb{R}^1$, then there exists a vector $f' \in \mathcal{D}(A_0)$ such that

$$\Phi(f', \lambda) = \frac{\Phi(f, \lambda)}{\lambda - \lambda_0} . \tag{8.5}$$

In fact, the condition (8.4) implies the existence of an element $f' \in \mathcal{D}(A_0)$ satisfying the equality

$$A_0 f' - \lambda_0 f' = f .$$

Applying the functional $\Phi(f, \lambda)$ to the two sides of this equality, we obtain, in view of the property a), the relation (8.5).

c) For every finite interval $\mathcal{I}$ of the real axis, an element $u \in \mathfrak{L}$ can be found such that

$$\Phi(u, \lambda) \not\equiv 0 \text{ as } \lambda \in \mathcal{I} . \tag{8.6}$$

According to the property 2) of a directing functional, there exist $\lambda \in \mathbb{R}^1$ and a vector $f_1 \in \mathfrak{L}$ such that

$$\Phi(f_1, \lambda) \neq 0 .$$

Suppose $\lambda_0 \in \mathcal{I}$ is a zero of the function $\Phi(f_1, \lambda)$,

$$\Phi(f_1, \lambda_0) = 0 .$$

Because of b), an element f'_1 can be found so that

$$\Phi(f'_1, \lambda) = \frac{\Phi(f_1, \lambda)}{\lambda - \lambda_0} . \tag{8.7}$$

Since the function $\Phi(f_1, \lambda)$ is analytic on the whole real axis, the interval $\mathcal{I}$ contains a finite number of zeroes of this function. Moreover, their multiplicity is finite. Denote by m the summarized multiplicity of all the zeroes. Repeating the procedure in (8.7) m times and taking into account that every such action reduces the multiplicity of a corresponding zero by 1, we find the desired vector u.

8.2   We shall call a linear set $\mathfrak{L}$ a quasi-Hilbert space if a sesquilinear form (f, g) (a so-called quasiscalar product) is given on it, which possesses all the properties of a scalar product except, possibly, $(f, f) > 0$ as $f \neq 0$. For a quasiscalar product the Schwartz inequality

$$|(f, g)|^2 \leq (f, f)(g, g), \quad f, g \in \mathfrak{L} , \tag{8.8}$$

holds true. Therefore the set $\hat{0} = \{g \in \mathfrak{L} : (g,g) = 0\}$ is linear. Denote by $\widehat{\mathfrak{L}}$ the factor space $\mathfrak{L}/\hat{0}$:

$$\widehat{\mathfrak{L}} = \mathfrak{L}/\hat{0}$$

and by $\hat{g}$ the element from $\widehat{\mathfrak{L}}$ which contains $g \in \mathfrak{L}$ (recall that elements of $\widehat{\mathfrak{L}}$ are classes of vectors from $\mathfrak{L}$ which differ from each other by an element belonging to $\hat{0}$). The scalar product in $\widehat{\mathfrak{L}}$ is introduced as

$$(\hat{f}, \hat{g}) = (f, g), \quad f \in \hat{f}, \, g \in \hat{g}.$$

It does not depend on the choice of vectors $f \in \hat{f}$ and $g \in \hat{g}$. Completing $\widehat{\mathfrak{L}}$ with respect to this scalar product, we obtain a certain Hilbert space denoted later on by $\mathfrak{H}_\mathfrak{L}$.

A linear operator A_0 on a quasi-Hilbert space $\mathfrak{L}$ is called Hermitian if $\mathcal{D}(A_0)$ is quasidense in $\mathfrak{L}$ and

$$(A_0 f, g) = (f, A_0 g), \quad f, g \in \mathcal{D}(A_0).$$

Due to (8.8), such an operator transforms the set $\{g_0 \in \mathcal{D}(A_0) : g_0 \in \hat{0}\}$ into $\hat{0}$. So, every Hermitian operator A_0 on $\mathfrak{L}$ generates a Hermitian operator $\widehat{A}$ on $\mathfrak{H}_\mathfrak{L}$ in the following way:

$$\widehat{A}\hat{g} = \widehat{A_0 g},$$
$$\mathcal{D}(\widehat{A}) = \{\hat{g} \in \widehat{\mathfrak{L}} : g \in \mathcal{D}(A_0)\}.$$

We denote by A the closure of $\widehat{A}$.

By the deficiency index of the operator A_0 on $\mathfrak{L}$ we mean that of the operator A on $\mathfrak{H}_\mathfrak{L}$ generated by A_0.

The operator A_0 is called positive if

$$(A_0 g, g) \geq 0, \quad g \in \mathcal{D}(A_0).$$

It is obvious that if A_0 is positive, then the operator A is positive too, and vice versa.

8.3 The following important theorem is valid.

Theorem 8.1 *Let A_0 be a Hermitian operator on a quasi-Hilbert space $\mathfrak{L}$, which has a directing functional $\Phi(f, \lambda)$. Then there exists at least one non-decreasing function*

$$\tau(\lambda) : \tau(0) = 0, \; \tau(\lambda - 0) = \tau(\lambda), \quad \lambda \in \mathbb{R}^1,$$

such that

$$(f, g) = \int_{-\infty}^{\infty} \Phi(f, \lambda)\overline{\Phi(g, \lambda)} \, d\tau(\lambda) \tag{8.9}$$

for arbitrary vectors $f, g \in \mathfrak{L}$. If the operator A_0 is positive, then the function $\tau(\lambda)$ in (8.9) can be chosen so that $\tau(\lambda) = 0$ when $\lambda \leq 0$.

Proof. Suppose first that $(f, f) > 0$ for any $f : 0 \neq f \in \mathfrak{L}$. In this case $\widehat{\mathfrak{L}} = \mathfrak{L}$, $\overline{\mathfrak{L}} = \mathfrak{H}_{\mathfrak{L}} = \mathfrak{H}$, and A is a Hermitian operator on the Hilbert space $\mathfrak{H}$.

Let $\mathcal{I}$ be a finite open interval of the real axis and u a vector from $\mathfrak{L}$ such that

$$\Phi(u, \lambda) \neq 0, \quad \lambda \in \mathcal{I} .$$

Such a vector exists due to property c). We assign to each vector $f \in \mathfrak{H}$ the function

$$F(\lambda) = \frac{\Phi(f, \lambda)}{\Phi(u, \lambda)}, \quad \lambda \in \mathcal{I} .$$

It is easily seen that

$$\Phi(f_\lambda, \lambda) = 0, \quad \lambda \in \mathcal{I}, \quad \text{where } f_\lambda = f - F(\lambda)u .$$

Hence, there exists a vector $g_\lambda \in \mathcal{D}(A)$ such that

$$f_\lambda = A g_\lambda - \lambda g_\lambda . \tag{8.10}$$

Let E_λ be a spectral function of the operator A, that is,

$$E_\lambda = P \widetilde{E}_\lambda ,$$

where $\widetilde{E}_\lambda$ is the spectral function of the corresponding self-adjoint extension, generally with exit into the larger space $\widetilde{\mathfrak{H}}$ of the operator A, and P is the orthoprojector from $\widetilde{\mathfrak{H}}$ onto $\mathfrak{H}$. Put

$$\Delta_\lambda^h = \begin{cases} \widetilde{E}_{\lambda+h} - \widetilde{E}_\lambda & \text{as} \quad h \leq 0 \\ \widetilde{E}_{\lambda+h} - \widetilde{E}_{\lambda+0} & \text{as} \quad h > 0. \end{cases}$$

Then

$$\lim_{h \to 0} \left\| \frac{1}{h} \left(\Delta_\lambda^h f - F(\lambda) \Delta_\lambda^h u \right) \right\| = \lim_{h \to 0} \left\| \frac{\Delta_\lambda^h f_\lambda}{h} \right\|$$

$$= \lim_{h \to 0} \left\| \frac{1}{h} \Delta_\lambda^h (A g_\lambda - \lambda g_\lambda) \right\| = \lim_{h \to 0} \frac{1}{h} \left| \int_{\substack{\lambda+0 \text{ if } h > 0 \\ \lambda-0 \text{ if } h \leq 0}}^{\lambda+h} (t - \lambda)^2 \, d(E_t g_\lambda, g_\lambda) \right|^{1/2}$$

$$= \lim_{h \to 0} \left| \int_{\substack{\lambda+0 \text{ if } h > 0 \\ \lambda-0 \text{ if } h \leq 0}}^{\lambda+h} d(E_t g_\lambda, g_\lambda) \right|^{1/2} = 0$$

$$\tag{8.11}$$

for any $f \in \mathfrak{L}$, any $\lambda \in \mathcal{I}$.

On the other hand,

$$\left\| \frac{1}{h}\left(F(\lambda)\Delta_\lambda^h u - \int_\lambda^{\lambda+h} F(t)\, d\widetilde{E}_t u \right) \right\|^2 = \left\| \frac{1}{h} \int_\lambda^{\lambda+h} (F(\lambda) - F(t))\, d\widetilde{E}_t u \right\|^2$$

$$= \frac{1}{h^2} \int_\lambda^{\lambda+h} |F(\lambda) - F(t)|^2 \, d\big(\widetilde{E}_t u\, , u\big) \le m^2 \Big((E_{t+h}u\, , u) - (E_{t+0}u\, , u) \Big)$$

if, for example, $h > 0$, where

$$m = \max_{\lambda \in [\lambda, \lambda+h]} |F'(\lambda)|\,.$$

It follows that

$$\lim_{h\to 0} \left\| \frac{1}{h}\left(F(\lambda)\Delta_\lambda^h u - \int_\lambda^{\lambda+h} F(t)\, d\widetilde{E}_t u \right) \right\| = 0\,. \tag{8.12}$$

The relations (8.11) and (8.12) lead to the equality

$$\lim_{h\to 0} \left\| \frac{1}{h}\left(\Delta_\lambda^h f - \int_\lambda^{\lambda+h} F(t)\, d\widetilde{E}_t u \right) \right\| = 0\,. \tag{8.13}$$

Thus, if a is a fixed point from the interval $\mathcal{I}$, then the vector-function

$$\varphi(\lambda) = \int_a^\lambda d\widetilde{E}_t f - \int_a^\lambda F(t)\, d\widetilde{E}_t u$$

has a derivative equal to 0 everywhere in $\mathcal{I}$. Since $\varphi(a) = 0$, $\varphi(\lambda) \equiv 0$ on $\mathcal{I}$. Consequently, for arbitrary $\lambda', \lambda'' \in \mathcal{I}$ and $f \in \mathfrak{L}$,

$$\widetilde{E}_{\lambda''} f - \widetilde{E}_{\lambda'} f = \int_{\lambda'}^{\lambda''} F(t)\, d\widetilde{E}_t u = \int_{\lambda'}^{\lambda''} \frac{\Phi(f,t)}{\Phi(u,t)}\, d\widetilde{E}_t u\,, \tag{8.14}$$

whence

$$\Big((E_{\lambda''} - E_{\lambda'})f, g \Big) = \Big((\widetilde{E}_{\lambda''} - \widetilde{E}_{\lambda'})f, (\widetilde{E}_{\lambda''} - \widetilde{E}_{\lambda'})g \Big)$$

$$= \int_{\lambda'}^{\lambda''} \frac{\Phi(f,t)\overline{\Phi(g,t)}}{|\Phi(u,t)|^2}\, d\big(E_t u\, , u\big) \tag{8.15}$$

as $f, g \in \mathfrak{L}$, $\lambda', \lambda'' \in \mathcal{I}$.

Now let $\lambda_0 \in \mathbb{R}^1$. Take an arbitrary finite interval containing the points 0 and λ_0 as $\mathcal{I}$, and a vector u satisfying (8.6), and set

$$\tau(\lambda_0) = \int\limits_0^{\lambda_0} \frac{1}{|\Phi(u,t)|^2} \, d(E_t u, u).$$

(8.16)

The function $\tau(\lambda_0)$, $\lambda_0 \in \mathcal{I}$, does not depend on the choice of u. Indeed, suppose that an element v possesses the property (8.6), i.e. $\Phi(v, \lambda) \neq 0$ for any $\lambda \in \mathcal{I}$. Using (8.15), we find that

$$\int\limits_0^{\lambda_0} \frac{1}{|\Phi(v,t)|^2} \, d(E_t v, v) = \int\limits_0^{\lambda_0} \frac{1}{|\Phi(v,t)|^2} \cdot \frac{|\Phi(v,t)|^2}{|\Phi(u,t)|^2} \, d(E_t u, u)$$

$$= \int\limits_0^{\lambda_0} \frac{1}{|\Phi(u,t)|^2} \, d(E_t u, u) \,,$$

hence, $\tau(\lambda_0)$ is independent of the choice of u. It follows from the definition (8.16) and the representation (8.15) that

$$\left(E_{\lambda''} f, g \right) - \left(E_{\lambda'} f, g \right) = \int\limits_{\lambda'}^{\lambda''} \Phi(f, \lambda)\overline{\Phi(g, \lambda)} \, d\tau(\lambda)$$

(8.17)

for $\lambda', \lambda'' \in \mathcal{I}$. Since the interval $\mathcal{I}$ may be chosen as large as desired, by the property c) of a directing functional, it is possible to pass to the limit in (8.17) when $\lambda' \to -\infty$, $\lambda'' \to \infty$. Thus we arrive at the representation (8.9).

As it is seen from the construction of $\tau(\lambda)$, this function does not decrease, and the normalization conditions $\tau(0) = 0$, $\tau(\lambda - 0) = \tau(\lambda)$ are satisfied.

So, the first assertion of the theorem is true under the assumption $\widehat{\mathfrak{L}} = \mathfrak{L}$. The second follows from the fact that every positive Hermitian operator admits a positive self-adjoint extension within $\mathfrak{H}$. The spectral function $\widetilde{E}_\lambda$ of such an extension possesses the property $\widetilde{E}_\lambda = 0$ as $\lambda < 0$. Then the function $\tau(\lambda)$ corresponding to this $\widetilde{E}_\lambda$ is the desired one because $\tau(\lambda) = 0$ when $\lambda < 0$.

To prove the theorem in the general situation when $\widehat{\mathfrak{L}} \neq \mathfrak{L}$, we replace the operator A in the foregoing arguments by $A = \widehat{\widetilde{A}}$, and mean by E_λ a spectral function of this operator A. Now we define the function $\tau(\lambda)$ by the formula (8.16) in which the element u is replaced by $\hat{u}$. Observing that (8.10) implies

$$\hat{f}_\lambda = A\hat{g}_\lambda - \lambda\hat{g}_\lambda, \quad \hat{g}_\lambda \in \mathcal{D}(A) \,,$$

and the equalities (8.11)–(8.13) are valid if the sign "^" stands over every vector from $\mathfrak{L}$, we conclude that

$$\tilde{E}_{\lambda''}\hat{f} - \tilde{E}_{\lambda'}\hat{f} = \int_{\lambda'}^{\lambda''} \frac{\Phi(f,\lambda)}{\Phi(u,\lambda)}\, d\tilde{E}_\lambda \hat{u}\,. \tag{8.18}$$

It is not difficult to verify that every equality succeeding (8.14) remains true if we replace the vectors from $\mathfrak{L}$ by the corresponding elements from $\widehat{\mathfrak{L}}$ at certain places but not everywhere, because the operator $\widehat{A}$ may have no directing functional. The point is that the identity $\Phi(f,\lambda) \equiv 0$ need not be fulfilled if $f \in \hat{0}$, since the functional $\Phi(f,\lambda)$ on $\mathfrak{L}$ does not yet generate a one-valued functional on $\widehat{\mathfrak{L}}$. Therefore the fact that the theorem is true in the case of $\widehat{\mathfrak{L}} = \mathfrak{L}$ does not at once imply its correctness in the general situation. However, continuing as specified above, we get representation (8.9), where $(\hat{f},\hat{g})$ stands for (f,g). Taking into account that $(\hat{f},\hat{g}) = (f,g)$, we obtain (8.9), which completes the proof.   □

8.4   Denote by $T(A_0)$ the set of all non-decreasing functions

$$\tau(\lambda) : \tau(0) = 0, \quad \tau(\lambda-0) = \tau(\lambda), \quad \lambda \in \mathbb{R}^1\,,$$

which yield the representation (8.9), i.e.

$$(f,g) = \int_{-\infty}^{\infty} \Phi(f,\lambda)\overline{\Phi(g,\lambda)}\, d\tau(\lambda) \quad f,g \in \mathfrak{L}\,.$$

We shall show presently that each $\tau(\lambda) \in T(A_0)$ can be obtained in the way mentioned when proving Theorem 8.1.
    Let $\tau(\lambda) \in T(A_0)$. Associate with a vector $f \in \mathfrak{L}$ the function

$$\varphi_f(\lambda) = \Phi(f,\lambda)\,.$$

In view of the isometry property (8.9), the map

$$f \mapsto \varphi_f(\lambda)$$

transforms $\widehat{\mathfrak{L}}$ onto a certain set $\widehat{\mathfrak{L}}_\tau$ of the Hilbert space $L_2(\mathbb{R}^1, d\tau)$. Moreover, the operator A is transformed into a certain restriction $\overset{\circ}{\Lambda}_\tau$ of the self-adjoint operator Λ_τ of multiplication by λ on $L_2(\mathbb{R}^1, d\tau)$. Denote by $\widehat{E}_t$ the spectral function of Λ_τ. Recall that $\widehat{E}_t$ acts as the multiplication by the characteristic function of the interval $(-\infty, t]$. Identifying f and $\varphi_f(\lambda)$, A and $\overset{\circ}{\Lambda}_\tau$, $\tilde{E}_t$ and $\widehat{E}_t$, we may consider the operator Λ_τ as a certain self-adjoint extension $\widehat{A}$ with exit into a space larger than $\mathfrak{H}_\mathfrak{L}$ of the operator A.
    Fix $\lambda_0 \in \mathbb{R}^1$, $\lambda_0 \neq 0$, arbitrarily, and suppose a vector $u \in \mathfrak{L}$ to be such that

$$\Phi(u,\lambda) \neq 0 \quad \text{as } \lambda \in [0,\lambda_0]\,.$$

Taking into account that the element u is identified with the function $\varphi_u(\lambda) = \Phi(u, \lambda)$, we get

$$\int_0^\lambda \frac{1}{|\Phi(u,t)|^2} \, d(E_t \hat{u}, \hat{u}) = \int_0^\lambda \frac{1}{|\Phi(u,t)|^2} \, d \int_{-\infty}^t |\Phi(u,s)|^2 \, d\tau(s) = \tau(\lambda), \quad \lambda \in [0, \lambda_0],$$

which proves, since $\lambda_0 \in \mathbb{R}^1$ is arbitrary, that $\tau(\lambda) \in T(A_0)$ is obtained in the desired way, that is by the formula

$$\tau(\lambda) = \int_0^\lambda \frac{1}{|\Phi(u,t)|^2} \, d(E_t \hat{u}, \hat{u}) . \tag{8.19}$$

The following theorem results from the foregoing arguments.

Theorem 8.2 *The set $T(A_0)$ coincides with the set of all the functions which are given by the formula (8.19), where E_t runs through the set of all spectral functions of the operator A.*

It is already not hard to prove the next assertion.

Theorem 8.3 *Under the conditions of Theorem 8.1, the function $\tau(\lambda) \in T(A_0)$ is unique if and only if at least one defect number of the operator A_0 is equal to zero.*

Proof. Let E_λ denote a spectral function of the operator A. As is known, it is uniquely determined by the operator A if and only if this operator is maximal. Therefore if E_λ is unique, then the function $\tau(\lambda) \in T(A_0)$ given by the formula (8.19) is also unique. Thus, the theorem will be proved if we show that different functions from $T(A_0)$ correspond to different spectral functions of A. So let the spectral functions E_λ^1 and E_λ^2 of the operator A be not identical. By virtue of $\overline{\mathfrak{L}} = \mathfrak{H}_\mathfrak{L}$, there exist elements $\hat{f}, \hat{g} \in \hat{L}$ such that

$$\left(E_\lambda^1 \hat{f}, \hat{g}\right) \neq \left(E_\lambda^2 \hat{f}, \hat{g}\right) .$$

But then in accordance with (8.15)

$$\int_{-\infty}^{\lambda_0} \Phi(f, \lambda)\overline{\Phi(g, \lambda)} \, d\tau_1(\lambda) \neq \int_{-\infty}^{\lambda_0} \Phi(f, \lambda)\overline{\Phi(g, \lambda)} \, d\tau_2(\lambda) .$$

Hence, the points $t \in \mathbb{R}^1$ can be found where $\tau_1(\lambda) \neq \tau_2(\lambda)$. $\qquad \square$

Theorem 8.4 *Suppose an operator A_0 on $\mathfrak{L}$ satisfies the conditions of Theorem 8.1. In addition, let a vector $v \in \mathfrak{L}$, $v \neq 0$, exist such that*

$$(v, v) = 0 . \tag{8.20}$$

Then $T(A_0)$ consists of only one function $\tau(\lambda)$. This function is constant on each interval which does not contain zeroes of the function $\Phi(v, \lambda)$. In this case the defect numbers of the operator A_0 equal 0, and its spectrum coincides with the set of those zeroes of the function $\Phi(v, \lambda)$ where the function $\tau(\lambda)$ has jumps.

Proof. Let E_λ be a spectral function of the operator A. Take an interval $\mathcal{I}_0 = (\lambda_0 - \delta, \lambda_0 + \delta)$ without zeroes of the function $\Phi(v, \lambda)$. Since (8.20) holds, the formula (8.18) yields

$$\widetilde{E}_{\lambda'} \hat{f} = \widetilde{E}_{\lambda''} \hat{f} \tag{8.21}$$

for all $f \in \mathcal{L}$ and λ', $\lambda'' \in \mathcal{I}_0$. Consequently,

$$\left\| A\hat{f} - \lambda_0 \hat{f} \right\|^2 = \int_{-\infty}^{\infty} (\lambda - \lambda_0)^2 \, d(E_\lambda \hat{f}, \hat{f})$$

$$= \left(\int_{-\infty}^{\lambda_0 - \delta} + \int_{\lambda_0 + \delta}^{\infty} \right) (\lambda - \lambda_0)^2 \, d(E_\lambda \hat{f}, \hat{f}) \geq \delta^2 \|f\|^2 \,.$$

As we can see, λ_0 is a point of regular type for the operator A, which is possible provided the defect numbers of A are equal. If we suppose that the deficiency index of A_0 is not $(0,0)$, then, by Proposition 1.3.5, the self-adjoint extension of the operator A can be found whose spectrum contains the point λ_0, contrary to the relation (8.21). Thus, the closure of $\widehat{A}$ is a self-adjoint operator, which implies the uniqueness of $\tau(\lambda) \in T(A_0)$. Moreover, (8.21) shows that the spectrum of the operator A consists of (not necessarily all) zeroes of the function $\Phi(v, \lambda)$. The equality (8.19) enables to conclude that $\tau(\lambda)$ is constant on every interval where there are no points of the spectrum of the operator A. It remains to be proved that if μ belongs to the spectrum of A, then $\tau(\mu + 0) - \tau(\mu) \neq 0$.

Let the interval $(\mu - \delta, \mu + \delta)$ have no points of the spectrum of the operator A different from μ. Then the function $\tau(\lambda)$ retains the same value inside $(\mu - \delta, \mu)$ as well as inside the interval $(\mu, \mu + \delta)$. Suppose $\tau(\mu + 0) = \tau(\mu)$. By the representation (8.9), for any $f \in \mathcal{D}(A_0)$

$$\left(A_0 f - \mu f, A_0 f - \mu f \right) = \int_{-\infty}^{\infty} (\lambda - \mu)^2 \left| \Phi(f, \lambda) \right|^2 d\tau(\lambda)$$

$$\geq \delta^2 \int_{-\infty}^{\infty} \left| \Phi(f, \lambda) \right|^2 d\tau(\lambda) = \delta^2 (f, f) \,.$$

In other words

$$\left\| A_0 \hat{f} - \mu \hat{f} \right\| \geq \delta \|\hat{f}\| \,,$$

contrary to the fact that the point μ belongs to the spectrum of A. $\square$

Theorem 8.5 *If a Hermitian operator A_0 on $\mathfrak{L}$ whose deficiency index is (1,1) has a directing functional, then*

$$(f, f) > 0 \quad \text{as} \quad f \neq 0,$$

and the operator A_0 is regular.

Proof. The non-singularity of the quasiscalar product $(\cdot, \cdot)$ in $\mathfrak{L}$ has been established in the previous theorem. We shall prove the regularity of the operator A_0, i.e. that every real number is a point of regular type for the operator A.

Let $0 \neq u \in \mathfrak{L}$ be the element such that the function $\Phi(u, \lambda) \not\equiv 0$. Denote by N_0 the set of all zeroes of the function $\Phi(u, \lambda)$ and associate with each vector $f \in \mathfrak{L}$ the vector-function

$$f_M(\lambda) = f_u(\lambda)u, \quad \lambda \notin N_0,$$

so that the equation

$$A_0 g - \lambda g = f - f_u(\lambda)u \tag{8.22}$$

has a solution $g \in \mathcal{D}(A_0)$. The equality

$$\Phi(f - f_u(\lambda)u, \lambda) = 0$$

is a necessary and sufficient condition for the solvability of the equation (8.22). It follows from here that the function

$$f_u(\lambda) = \frac{\Phi(f, \lambda)}{\Phi(u, \lambda)}$$

is uniquely determined. Since the functions $\Phi(f, \lambda)$ and $\Phi(u, \lambda)$ are analytic on the real axis, for any $\lambda_0 \in \mathbb{R}^1$ there exists $p \geq 0$ such that the function

$$(\lambda - \lambda_0)^p f_u(\lambda)$$

is analytic in a sufficiently small neighbourhood of λ_0. Consequently, the conditions of Theorem 3.3 are satisfied. Therefore the whole real axis consists of the points of regular type of the operator A. $\qquad\square$

8.5   The following class of directing functionals is of great importance.

Definition 8.2 *We call a directing functional universal if for every fixed $f \in \mathfrak{L}$ the function $\Phi(f, \lambda)$ is entire and the condition*

$$\Phi(f, z) = 0, \quad z \in \mathbb{C}^1,$$

is necessary and sufficient for the equation

$$A_0 g - zg = f$$

to be solvable.

If a Hermitian operator A_0 has a universal directing functional, then according to the definition, the dimension of the factor space

$$\mathcal{L}/(A_0 - zI)\mathcal{D}(A_0)$$

is not larger than 1. This implies that the defect numbers of A_0 do not exceed 1. We shall be interested only in the case where the deficiency index of A_0 is (1,1).

Suppose now that the universal directing functional $\Phi(f, z)$ possesses the property

$$\Phi(u, z) = 1 \tag{8.23}$$

for some vector $u \in \mathfrak{H}$. Choose this vector as a gauge in the isomorphism (1.2.6). Then

$$f_u(z) = \Phi(f, z).$$

So the operator A is entire. The function $\tau(\lambda)$ giving the representation (8.9) of the scalar product (f, g) is bounded. In this case we may normalize it by the conditions

$$\tau(-\infty) = 0, \quad \tau(\lambda - 0) = \tau(\lambda).$$

The function $\tau(\lambda)$ normalized by these conditions coincides with a certain distribution function associated with the gauge u of the operator A, that is

$$\tau(\lambda) = (E_\lambda u, u),$$

where E_λ is a spectral function of the operator A.

9　Operators, Entire with Respect to a Generalized Gauge

In this section A, $\overline{\mathcal{D}(A)} = \mathfrak{H}$, is a closed simple Hermitian operator in $\mathfrak{H}$ whose deficiency index is (1,1).

9.1　We start with the definition.

Definition 9.1 *We shall call a linear functional $\tilde{u}$ given on $\mathcal{D}(A^*)$ a generalized element (vector) if there exists a sequence $\{u_n\}_{n \in \mathbb{N}}$, $u_n \in \mathfrak{H}$, such that*

$$(f, u_n) \to \tilde{u}(f), \quad n \to \infty$$

for every $f \in \mathcal{D}(A^*)$ ($\tilde{u}(\cdot)$ means the action of the functional $\tilde{u}$).

On the set $\mathcal{D}(A^*)$ we introduce the scalar product

$$(f, g)_+ = (f, g) + (A^*f, A^*g), \quad f, g \in \mathcal{D}(A^*). \tag{9.1}$$

As has been mentioned in subsection 1.1.1, $\mathcal{D}(A^*)$ is a complete Hilbert space with respect to the scalar product (9.1). Denote it by $\mathfrak{H}_+$.

It is easily checked that

$$\|f\| \leq \|f\|_+, \quad f \in \mathfrak{H}_+, \tag{9.2}$$

that is $\mathfrak{H}_+$ is a space with positive norm in relation to $\mathfrak{H}$ in the sense of Berezansky [1].

Every vector $f \in \mathfrak{H}$ generates the linear functional l_f on $\mathfrak{H}_+$ if we put

$$l_f(g) = (g, f), \quad g \in \mathfrak{H}_+.$$

The functional l_f is continuous because, in view of (9.2),

$$|l_f(g)| = |(g, f)| \leq \|g\| \cdot \|f\| \leq \|f\| \cdot \|g\|_+,$$

whence

$$\|l_f\| = \sup_{g \in \mathfrak{H}_+} \frac{|(g, f)|}{\|g\|_+} \leq \|f\|.$$

The completion of $\mathfrak{H}$ in the norm

$$\|f\|_- = \|l_f\|$$

denoted by $\mathfrak{H}_-$ is a space with negative norm constructed from the spaces $\mathfrak{H}$ and $\mathfrak{H}_+$. So, we have the chain of Hilbert spaces

$$\mathfrak{H}_+ \subseteq \mathfrak{H} \subseteq \mathfrak{H}_-$$

densely and continuously imbedded into each other. Obviously, the form (g, f) given on $\mathfrak{H}_+ \times \mathfrak{H}$ can be extended continuously to $\mathfrak{H}_+ \times \mathfrak{H}_-$. In what follows, (g, f), $g \in \mathfrak{H}_+$, $f \in \mathfrak{H}_-$ means the value of the extended form.

Since, in accordance with (6.11), a generalized resolvent R_z of the operator A ($R_z = P\tilde{R}_z$, $\Im z \neq 0$, $\tilde{R}_z$ is the resolvent of a corresponding self-adjoint extension $\tilde{A}$, generally, with exit into a larger space $\tilde{\mathfrak{H}} \supset \mathfrak{H}$, P is the orthogonal projector from $\tilde{\mathfrak{H}}$ onto $\mathfrak{H}$) transforms the space $\mathfrak{H}$ into $\mathcal{D}(A^*)$ and

$$(A^* - zI)R_z f = f, \quad f \in \mathfrak{H},$$

R_z is a continuous linear operator from $\mathfrak{H}$ into $\mathfrak{H}_+$. Therefore the adjoint $\widehat{R}_z$ of the operator R_z, which is defined by the relation

$$(R_z f, g) = (f, \widehat{R}_{\bar{z}} g), \quad f \in \mathfrak{H}, \quad g \in \mathfrak{H}_-,$$

acts continuously from $\mathfrak{H}_-$ into $\mathfrak{H}$. This is an extension of $R_{\bar{z}}$ from $\mathfrak{H}$ to $\mathfrak{H}_-$:

$$\widehat{R}_z f = R_{\bar{z}} f \quad \text{as } f \in \mathfrak{H}.$$

Theorem 9.1 *A linear functional $\tilde{u}$ on $\mathcal{D}(A^*)$ is a generalized element if and only if $\tilde{u} \in \mathfrak{H}_-$.*

Proof. If $\tilde{u}$ is a generalized element, then $\tilde{u}$ is the weak limit of a certain sequence $u_n \in \mathfrak{H} \subset \mathfrak{H}_-$. By virtue of the completeness of $\mathfrak{H}_-$ with respect to the weak convergence, $\tilde{u} \in \mathfrak{H}_-$. Conversely, if $\tilde{u} \in \mathfrak{H}_-$, then there exists a sequence $u_n \in \mathfrak{H}$ such that $u_n \to u$ in the space $\mathfrak{H}_-$ when $n \to \infty$. So, for any $f \in \mathfrak{H}_+$,

$$(f, u_n) \to (f, \tilde{u}),$$

which completes the proof. $\square$

Since the operator $\hat{R}_z$, $\Im z \neq 0$, acts continuously from $\mathfrak{H}_-$ into $\mathfrak{H}$ and coincides with $R_{\bar{z}}$ on $\mathfrak{H}$, the convergence $u_n \to \tilde{u}$ in $\mathfrak{H}_-$ implies the strong convergence in $\mathfrak{H}$ of the sequence $\overset{\circ}{R_{\bar{z}}} u_n$ to a certain element. We denote this element by $\hat{R}_z \tilde{u}$.

Now let $\overset{\circ}{A}$ be a self-adjoint extension within $\mathfrak{H}$ of the operator A, and $\overset{\circ}{R_z}$ its resolvent. Applying the Hilbert resolvent identity

$$\overset{\circ}{R_z} = \overset{\circ}{R_{z_0}} + (z - z_0)\overset{\circ}{R_z}\overset{\circ}{R_{z_0}} \tag{9.3}$$

to the elements u_n and then passing to the limit when $n \to \infty$, we get

$$\widehat{\overset{\circ}{R_z}} \tilde{u} = \widehat{\overset{\circ}{R_{z_0}}} \tilde{u} + (z - z_0)\overset{\circ}{R_z}\widehat{\overset{\circ}{R_{z_0}}} \tilde{u} . \tag{9.4}$$

The equality (9.4) shows that the vector-function $\widehat{\overset{\circ}{R_z}} \tilde{u}$ is analytic at every regular point of the operator $\overset{\circ}{A}$. On the other hand, writing the formula (6.20) for the vectors u_n, and passing to the limit as $n \to \infty$, we arrive at the equality

$$\widehat{R_z} \tilde{u} = \widehat{\overset{\circ}{R_z}} \tilde{u} + \frac{(\tilde{u}, \varphi(\bar{z}))\varphi(z)}{\tau(z) + Q_1(z)}, \tag{9.5}$$

valid for an arbitrary generalized resolvent R_z of the operator A; here the functions $\tau(z)$ and $Q_1(z)$ are the same as those in (6.20),

$$(\tilde{u}, \varphi(\bar{z})) = \overline{\tilde{u}(\varphi(\bar{z}))} . $$

It follows from (9.5) that for any generalized resolvent R_z of the operator A, the vector-function $\widehat{R_z} \tilde{u}$ taking its values in $\mathfrak{H}$ is analytic in the half-plane $\Im z > 0$.

9.2 Let the points z_0, z, z_1 be regular for the extension $\overset{\circ}{A}$. Applying the identity (9.3) twice to the resolvent $\overset{\circ}{R_z}$, we obtain

$$\overset{\circ}{R_z} = \overset{\circ}{R_{z_0}} + (z - z_0)\overset{\circ}{R_{z_1}}\overset{\circ}{R_{z_0}} + (z - z_0)(z - z_1)\overset{\circ}{R_z}\overset{\circ}{R_{z_1}}\overset{\circ}{R_{z_0}} . \tag{9.6}$$

Substituting in (9.6) $-i$ and i for z_1 and z_0, respectively and vice versa, for z_0 and z_1, and then adding, we find that

$$\overset{\circ}{R_z} - \frac{\overset{\circ}{R_i} + \overset{\circ}{R_{-i}}}{2} = z\overset{\circ}{R_{-i}}\overset{\circ}{R_i} + (z - i)(z + i)\overset{\circ}{R_z}\overset{\circ}{R_i}\overset{\circ}{R_{-i}} . \tag{9.7}$$

If $\widetilde{u}$ is a generalized element, then the sequence of the R-functions $\left(\mathring{R}_z u_n, u_n\right)$ does not generally converge as $n \to \infty$. However, if $\left(\mathring{R}_z u_n, u_n\right)$ converges at a point $z_0 : \Im z_0 \neq 0$, then, as follows from (9.6), this sequence converges at any other point $z : \Im z \neq 0$. Its limit is an R-function too. The generalized resolvent formula (6.20) shows that in this case the sequence $\left(R_z u_n, u_n\right)$ has a limit in the class of R-functions, where R_z is an arbitrary generalized resolvent of the operator A.

If the sequence $\left(\mathring{R}_z u_n, u_n\right)$ is not converging at any point $z : \Im z \neq 0$, then we consider the regularized sequence

$$\left(\left(\mathring{R}_z - \frac{\mathring{R}_i + \mathring{R}_{-i}}{2}\right)u_n, u_n\right).$$

In view of (9.7), this sequence has a limit. If we denote this limit by $\langle\widehat{\mathring{R}}_z\widetilde{u}, \widetilde{u}\rangle$, then

$$\langle\widehat{\mathring{R}}_z\widetilde{u}, \widetilde{u}\rangle = z\big(\widehat{\mathring{R}}_i\widetilde{u}, \widehat{\mathring{R}}_i\widetilde{u}\big) + (z - i)(z + i)\big(R_z\widehat{\mathring{R}}_i\widetilde{u}, \widehat{\mathring{R}}_i\widetilde{u}\big). \tag{9.8}$$

The application of (6.20) to the points z, i and $-i$ yields

$$\left(\left(R_z - \frac{R_i + R_{-i}}{2}\right)u_n, u_n\right)$$

$$= \left(\left(\mathring{R}_z - \frac{\mathring{R}_i + \mathring{R}_{-i}}{2}\right)u_n, u_n\right) + \frac{\left(u_n, \varphi(\bar{z})\right)\left(\varphi(z), u_n\right)}{\tau(z) + Q_1(z)} \tag{9.9}$$

$$- \frac{1}{2}\left(\frac{\left(u_n, \varphi(-i)\right)\left(\varphi(i), u_n\right)}{\tau(i) + Q_1(i)} + \frac{\left(u_n, \varphi(i)\right)\left(\varphi(-i), u_n\right)}{\tau(-i) + Q_1(-i)}\right).$$

The sequence in the right-hand side of (9.9) converges as $n \to \infty$. Denote its limit by $\langle\widehat{R}_z\widetilde{u}, \widetilde{u}\rangle$. Setting

$$\gamma(\widetilde{A}) = -\frac{1}{2}\left(\frac{\left(\widetilde{u}, \varphi(-i)\right)\left(\varphi(i), \widetilde{u}\right)}{\tau(i) + Q_1(i)} + \frac{\left(\widetilde{u}, \varphi(i)\right)\left(\varphi(-i), \widetilde{u}\right)}{\tau(-i) + Q_1(-i)}\right)$$

and passing to the limit in (9.9), we arrive at the relation

$$\langle\widehat{R}_z\widetilde{u}, \widetilde{u}\rangle = \langle\widehat{\mathring{R}}_z\widetilde{u}, \widetilde{u}\rangle + \frac{\left(\widetilde{u}, \varphi(\bar{z})\right)\left(\varphi(z), \widetilde{u}\right)}{\tau(z) + Q_1(z)} + \gamma(\widetilde{A}); \tag{9.10}$$

here the constant $\gamma(\widetilde{A})$ depends on the choice of a generalized resolvent R_z. Reducing to a common denominator and changing $\tau(\lambda)$ for $-\frac{1}{\tau(\lambda)}$ in (9.10), we obtain

$$w_{\widetilde{u}}(z) = \langle\widehat{R}_z\widetilde{u}, \widetilde{u}\rangle = \frac{p_1(z)\tau(z) + p_0(z)}{q_1(z)\tau(z) + q_0(z)} + \gamma(\widetilde{A}), \tag{9.11}$$

where $\tau(z)$ is an R-function,

$$q_0(z) = -\frac{1}{(\tilde{u}, \varphi(\bar{z}))} \; ;$$

$$q_1(z) = -Q_1(z)q_0(z) \; ;$$

$$p_0(z) = \overset{\circ}{w}_{\tilde{u}}(z)q_0(z) \; ;$$

$$p_1(z) = \overset{\circ}{w}_{\tilde{u}}(z)q_1(z) + (\varphi(z), \tilde{u}) \; ;$$

$$\overset{\circ}{w}_{\tilde{u}}(z) = \langle \widehat{R}_z \tilde{u}, \tilde{u} \rangle \; ;$$

$$Q_1(z) = i\Im z_0 + (z - z_0)(\varphi(z_0), \varphi(\bar{z})) \; ,$$

$\langle \widehat{R}_z \tilde{u}, \tilde{u} \rangle$ is defined by (9.8). The equality (9.11) may be considered as a generalization of the generalized resolvent formula (7.3) to the case where the gauge is a generalized element.

Theorem 9.2 *The function* $w_{\tilde{u}} = \langle \widehat{R}_z \tilde{u}, \tilde{u} \rangle$ *admits a representation of the form*

$$w_{\tilde{u}}(z) = \int\limits_{-\infty}^{\infty} \frac{1 + \lambda z}{\lambda - z} \, d\sigma(\lambda) \; , \tag{9.12}$$

where

$$\sigma(\lambda) = (E_\lambda \widehat{R}_i \tilde{u}, \widehat{R}_i \tilde{u}) \; ,$$

E_λ *is a spectral function of the operator* A.

Proof. Let $\{u_n\}_{n \in \mathbb{N}}$ be the sequence from Definition 9.1. Then

$$w_{\tilde{u}}(z) = \lim_{n \to \infty} \left((R_z u_n, u_n) - \left(\frac{R_i + R_{-i}}{2} u_n, u_n \right) \right)$$

$$= \lim_{n \to \infty} \int\limits_{-\infty}^{\infty} \left(\frac{1}{\lambda - z} - \frac{\lambda}{1 + \lambda^2} \right) d(E_\lambda u_n, u_n) \tag{9.13}$$

$$= \lim_{n \to \infty} \int\limits_{-\infty}^{\infty} \frac{1 + \lambda z}{\lambda - z} \cdot \frac{1}{1 + \lambda^2} \, d\rho_n(\lambda) \; ,$$

where

$$\rho_n(\lambda) = (E_\lambda u_n, u_n) \; .$$

Let $\widetilde{E}_\lambda$ denote the spectral function of the self-adjoint extension $\widetilde{A}$ of the operator A, possibly with exit to a larger space, which corresponds to E_λ, and $\widetilde{R}_z = (\widetilde{A} - \lambda I)^{-1}$. It follows from the equalities

$$(\widetilde{E}_\Delta \widetilde{R}_i u_n, \widetilde{R}_i u_n) = \int\limits_\Delta d(\widetilde{E}_\lambda \widetilde{R}_i u_n, \widetilde{R}_i u_n) = \int\limits_\Delta \frac{d\rho(\lambda)}{1 + \lambda^2}$$

that

$$\frac{d\rho(\lambda)}{1+\lambda^2} = d\big(\tilde{E}_\lambda \tilde{R}_i u_n, \tilde{R}_i u_n\big) .$$

Put

$$\sigma_n(\lambda) = \big(\tilde{E}_\lambda \tilde{R}_i u_n, \tilde{R}_i u_n\big) .$$

Since the sequence of functions $\sigma_n(\lambda)$ whose variations are bounded converges to $\sigma(\lambda)$, we can pass to the limit in (9.13) as $n \to \infty$ and obtain the desired result.□

Adding and subtracting z^2 in the numerator of the integrand in (9.12), we get another form of the representation of the function $w_{\tilde{u}}(z)$, namely:

$$w_{\tilde{u}}(z) = (1 + z^2) \int_{-\infty}^{\infty} \frac{d\sigma(\lambda)}{\lambda - z} + z \int_{-\infty}^{\infty} d\sigma(\lambda)$$

$$= z \|\tilde{R}_i \tilde{u}\|^2 + (1 + z^2) \int_{-\infty}^{\infty} \frac{d\sigma(\lambda)}{\lambda - z} .$$

(9.14)

The equality (9.14) shows that $w_{\tilde{u}}(z)$ is an N-function in the half-plane $\Im z > 0$.

9.3 Assume that for a given generalized element $\tilde{u}$ there exist points $z_+ : \Im z_+ > 0$ and $z_- : \Im z_- < 0$ such that

$$\big(\tilde{u}, \varphi(z_+)\big) \neq 0, \quad \big(\tilde{u}, \varphi(z_-)\big) \neq 0 ,$$

where

$$\varphi(z) = \varphi(z_0) + (z - z_0)\overset{\circ}{R}_z \varphi(z_0), \quad \varphi(z_0) \in \mathfrak{N}_{z_0} .$$

(9.15)

Since the function $\overset{\circ}{\widehat{R_z \tilde{u}}}$, $\Im z \neq 0$, is analytic, the set of all non-real zeroes of the function $\big(\tilde{u}, \varphi(\bar{z})\big)$ is at most countable. Therefore the function

$$f_{\tilde{u}}(z) = \frac{\big(f, \varphi(\bar{z})\big)}{\big(\tilde{u}, \varphi(\bar{z})\big)}$$

(9.16)

is meromorphic in the upper and lower half-planes of the complex plane. Taking into account that the functional $f - f_{\tilde{u}}(z)\tilde{u}$ vanishes on the space $\mathfrak{N}_{\bar{z}}$, we conclude that the function $f_{\tilde{u}}(z)$ defined by (9.16) does not depend on the choice of a canonical resolvent $\overset{\circ}{R}_z$ of the operator A in (9.15).

Denote by $\mathfrak{H}_{\tilde{u}}$ the set of the functions $f_{\tilde{u}}(z)$ assigned to all the elements $f \in \mathfrak{H}$ by the formula (9.16). Using a proof analogous to that of Theorem 1.2.2, one can prove that the mapping

$$f \mapsto f_{\tilde{u}}(z)$$

(9.17)

is a linear isomorphism from $\mathfrak{H}$ onto $\mathfrak{H}_{\tilde{u}}$. This isomorphism transforms the operator A into the operator of multiplication by the independent variable z. If $f_{\tilde{u}}(z) \in \mathfrak{H}_{\tilde{u}}$ and $f_{\tilde{u}}(a) = 0$, then the function

$$\frac{f_{\tilde{u}}(z)}{z - a}$$

also belongs to $\mathfrak{H}_{\tilde{u}}$.

As is seen from (9.16), the map (9.17) can be extended from $\mathfrak{H}$ to $\mathfrak{H}_-$. Denote by $\mathfrak{H}_{\tilde{u}}^-$ the image of this map:

$$\mathfrak{H}_{\tilde{u}}^- = \left\{ f_{\tilde{u}}(z) = \frac{(f, \varphi(\bar{z}))}{(\tilde{u}, \varphi(\bar{z}))}, \quad f \in \mathfrak{H}_- \right\}.$$

As $\tilde{u}_{\tilde{u}}(z) \equiv 1$, $1 \in \mathfrak{H}_{\tilde{u}}^-$.

If $\tilde{R}_z = (\tilde{A} - zI)^{-1}$, where $\tilde{A}$ is a self-adjoint extension of A, possibly, with exit to a certain larger space $\tilde{\mathfrak{H}}$, and $\psi_0 = \varphi(z_1) \in \mathfrak{N}_{z_1}$, then, generally, the values of the vector-function

$$\psi(z) = \psi_0 + (z - z_1)\tilde{R}_z\psi_0$$

may not belong to $\mathfrak{H}$. But the vector-function $P\psi(z)$ analytic on the resolvent set of the operator $\tilde{A}$ takes its values in the subspace $\mathfrak{N}_z$, i.e. $P\psi(z)$ is orthogonal to $\mathfrak{M}_{\bar{z}}$, whence

$$(f, \psi(\bar{z})) = f_{\tilde{u}}(z)(\tilde{u}, \psi(\bar{z})), \quad f \in \mathfrak{H}. \tag{9.18}$$

Substituting

$$\psi(\bar{z}) = \varphi(z_1) + (\bar{z} - z_1)\tilde{R}_{\bar{z}}\varphi(z_1) \tag{9.19}$$

into the left-hand side of equation (9.18) and

$$\psi(\bar{z}) = \varphi(z_1) + (\bar{z} - z_1)\tilde{R}_i\varphi(z_1) + (\bar{z} - z_1)(\bar{z} - i)\tilde{R}_{\bar{z}}\tilde{R}_i\varphi(z_1) \tag{9.20}$$

into its right-hand side (it follows from (1.1.2) that (9.19) and (9.20) are equivalent), and then multiplying the two sides by $(z - z_1)^{-1}$, and integrating along the broken path Δ_ε as in Theorem 1.2.5, we obtain the following assertion.

Theorem 9.3 *Let the function $f_{\tilde{u}}(z)$ assigned to $f \in \mathfrak{H}$ be analytic on the interval $[a, b] \subset \mathbb{R}^1$. Then*

$$E_\Delta f = \int_\Delta f_{\tilde{u}}(\lambda)(\lambda - i)\, dE_\lambda \hat{R}_i \tilde{u}, \tag{9.21}$$

where E_λ is any spectral function of the operator A.

The next statement is a consequence of (9.21).

Corollary 9.1 *Let E_λ be a spectral function of the operator A and*

$$d\sigma(\lambda) = |\lambda - i|^2 d\big(E_\lambda \widehat{R}_i \widetilde{u}, \widehat{R}_i \widetilde{u}\big) \,. \tag{9.22}$$

If for the elements $f, g \in \mathfrak{H}$ the functions $f_{\widetilde{u}}(z)$ and $g_{\widetilde{u}}(z)$ are analytic on the interval $[a, b]$, then

$$\big(E_\Delta f, g\big) = \int\limits_\Delta f_{\widetilde{u}}(\lambda) \overline{g_{\widetilde{u}}(\lambda)} \, d\sigma(\lambda) \,. \tag{9.23}$$

Denote by $V_{\widetilde{u}}(A)$ the set of functions $\sigma(\lambda)$ given by the formula (9.22) where E_λ runs through the set of all spectral functions of A. The function $\sigma(\lambda) \in V_{\widetilde{u}}(A)$ is called canonical if it is generated by an orthogonal spectral function E_λ. Let $\mathfrak{R}$ denote the set of $f \in \mathfrak{H}$ whose functions $f_{\widetilde{u}}(z)$ are analytic on the whole real axis.

We say that a generalized gauge $\widetilde{u}$ is quasiregular if $\overline{\mathfrak{R}} = \mathfrak{H}$. In what follows in this section we consider only a quasiregular gauge $\widetilde{u}$. Reasoning as in the proof of Theorems 1.1 and 1.2, we obtain, by using (9.23), the analogues of these theorems for the case of a generalized quasiregular gauge.

Theorem 9.4 *The set $V_{\widetilde{u}}(A)$ consists of those and only those non-decreasing functions*

$$\sigma(\lambda) : \int\limits_{-\infty}^{\infty} \frac{d\sigma(\lambda)}{1 + \lambda^2} < \infty$$

which yield the representation

$$(f, g) = \int\limits_{-\infty}^{\infty} f_{\widetilde{u}}(\lambda) \overline{g_{\widetilde{u}}(\lambda)} \, d\sigma(\lambda) \tag{9.24}$$

for any $f, g \in \mathfrak{R}$.

Theorem 9.5 *Let $\sigma(\lambda) \in V_{\widetilde{u}}(A)$. The mapping $f \mapsto f_{\widetilde{u}}(\lambda)$ from $\mathfrak{H}$ into $L_2\big(\mathbb{R}^1, d\sigma\big)$ is isometric. It is an isomorphism between $\mathfrak{H}$ and the whole space $L_2\big(\mathbb{R}^1, d\sigma\big)$ if and only if the function $\sigma(\lambda)$ is canonical.*

9.4 Using the formulas (9.20) and (9.24), one can show as in section 2 that $\big(\widetilde{u}, \varphi(\bar{z})\big)$ and $\big(f, \varphi(\bar{z})\big)$ are N-functions in the open upper and lower half-planes, so all the results of that section hold true in the case of a generalized gauge $\widetilde{u}$.

A point a is called $\widetilde{u}$-regular for the operator A if every function $f_{\widetilde{u}}(z) \in \mathfrak{H}_{\widetilde{u}}$ is analytic in some disk $K(a, r) : |z - a| < r$. Doing as in subsection 3.1, one can establish the following theorem, analogous to Theorem 3.1.

Theorem 9.6 *In order that a point a be $\widetilde{u}$-regular for the operator A, it is necessary and sufficient that it be of regular type for this operator and $(\widetilde{u}, \varphi(a)) \neq 0$.*

Using the arguments of subsection 3.2, it is also possible to prove the analogues of Theorems 3.2 and 3.3 in which the word "u-regular" is replaced by "$\widetilde{u}$-regular".

Now fix a $\widetilde{u}$-regular point z of A. The linear functional $f_{\widetilde{u}}(z)$ is continuous in $\mathfrak{H}$. Hence,

$$f_{\widetilde{u}}(z) = (f, \widetilde{e}(\bar{z})), \quad f \in \mathfrak{H}.$$

The value of the vector-function $\widetilde{e}(z)$ at the point z belongs to the subspace $\mathfrak{N}_z$. This function is analytic on the mirror image with respect to the real axis of the set of all $\widetilde{u}$-regular points of the operator A.

Set

$$\nabla_{\widetilde{u}}(z) = \sup_{f \in \mathfrak{H}} \frac{|f_{\widetilde{u}}(z)|}{\|f\|}.$$

All the results of section 4 concerning the functions $e(z)$ and $\nabla_u(z)$ may literally be carried over to the functions $\widetilde{e}(z)$ and $\nabla_{\widetilde{u}}(z)$.

Definition 9.2 *We shall call a closed simple Hermitian operator whose deficiency index is (1,1) entire with a generalized gauge if there exists a generalized element (gauge) $\widetilde{u}$ such that for any vector $f \in \mathfrak{H}$ the function $f_{\widetilde{u}}(z)$ is entire.*

Each point of the complex plane is $\widetilde{u}$-regular for an entire operator with a generalized gauge $\widetilde{u}$. It should also be noted that all principal results of sections 6–8 hold true for entire operators with a generalized gauge. We shall not deal with them in greater detail.

9.5   Let the operator A have a universal directing functional $\Phi(f, z)$ given also on some generalized element $\widetilde{u}$. Suppose that

$$\Phi(\widetilde{u}, z) \equiv 1.$$

In this case

$$f_{\widetilde{u}}(z) = \Phi(f, z).$$

If the deficiency index of A is $(1,1)$, then this operator is entire with a generalized gauge $\widetilde{u}$. The set of all functions $\tau(\lambda)$ giving the representation (8.9) coincides with the set of functions of the form (9.22) where E_λ goes through the set of all spectral functions of the operator A.

Chapter 3

Applications of Entire Operator Theory
to Some Classical Problems of Analysis

In this chapter the theory of entire operators developed in Chapter 2 is applied to some classical problems of analysis. The power moment problem is among them. This problem is considered in Section 1. It is shown there that a certain Hermitian operator on the appropriate Hilbert space can be associated with it. This operator has a directing functional. It is established that it is entire if and only if the moment problem under consideration is indefinite, i.e. nonuniquely solved. The indefiniteness criteria are found on the basis of the general theory of entire operators presented in the previous chapter. In the case where the problem is indefinite, all its solutions are effectively described. Section 2 deals with positive definite functions. The analogue of the classical Bochner theorem concerning an integral representation of a function positive definite on the whole real axis is proved for a positive definite function given on a finite interval. It turns out that the integral representation of the Bochner type for such a function can be obtained as a special case of expansion in eigenfunctions of a certain Hermitian operator having a directing functional. As opposed to the case of the whole real axis, this representation may be nonunique. In the indefinite case the operator whose deficiency index is (1,1) associated with the given function is entire, and the problem of finding all spectral functions of this operator or all the representations of the given positive definite function, which is the same, is equivalent to the description of all its positive definite continuations to the whole real axis. This problem is constructively solved by using the general theory of entire operators. Both definite and indefinite cases are illustrated by a number of useful examples. The representation and continuation problems for spiral and unitary arcs in a Hilbert space are discussed in section 3. In the case where a continuation of a spiral or unitary arc to a complete spiral or unitary line, respectively, is not unique, all the continuations are described. In contrast to the above problems, the Hermitian operator corresponding to the indefinite continuation problem for a spiral arc is entire with respect to a generalized not ordinary gauge. In this connection the general theory of such operators sketched in section 2.9 is used for solving the problems being considered.

1   Power Moment Problem

1.1   Let $\{s_k\}_{k\in\mathbb{N}_0}$, $\mathbb{N}_0 = \mathbb{N} \cup \{0\}$, be a sequence of numbers $s_k \in \mathbb{C}^1$ (without loss of generality we may put $s_0 = 1$). We shall be interested in the conditions under which the numbers s_k are the power moments of a certain distribution function $\sigma(\lambda)$, $\lambda \in \mathbb{R}^1$, that is, the sequence $\{s_k\}_{k\in\mathbb{N}_0}$ admits a representation of the form

$$s_k = \int_{-\infty}^{\infty} \lambda^k \, d\sigma(\lambda), \quad k \in \mathbb{N}_0 , \tag{1.1}$$

where $\sigma(\lambda)$, $\lambda \in \mathbb{R}^1$, is a bounded non-decreasing function normalized by the conditions

$$\sigma(\lambda - 0) = \sigma(\lambda), \ \sigma(-\infty) = 0 .$$

For a sequence $\{s_k\}_{k\in\mathbb{N}_0}$ which can be represented in the form (1.1), it is reasonable to ask when the representation (1.1) is unique. In the case of nonuniqueness the question arises of describing all the solutions of the power moment problem.

The solvability criterion of the moment problem (1.1) is simply established.

Theorem 1.1 *In order that a sequence* $\{s_k\}_{k\in\mathbb{N}_0}$, $s_k \in \mathbb{C}^1$, *admit a representation of the form* (1.1), *it is necessary and sufficient that*

$$\sum_{j,k=0}^{n} s_{j+k}\xi_j\bar{\xi}_k \geq 0, \quad n \in \mathbb{N}_0 , \tag{1.2}$$

for arbitrary numbers $\xi_j \in \mathbb{C}^1$, $j \in \mathbb{N}_0$.

Proof. The necessity immediately follows from the representation (1.1). Let us prove the sufficiency.

Denote by $\mathfrak{L}$ the linear set of all polynomials

$$f(\lambda) = \sum_{k=0}^{n} \xi_k\lambda^k , \quad \xi_k \in \mathbb{C}^1, \quad k \in \mathbb{N}_0 , \tag{1.3}$$

and set

$$(f,g) = \sum_{j=0}^{n}\sum_{k=0}^{m} s_{j+k}\xi_j\bar{\eta}_k \tag{1.4}$$

for f of the form (1.3) and

$$g(\lambda) = \sum_{0}^{m} \eta_k\lambda^k \quad \eta_k \in \mathbb{C}^1, \quad k = 0, 1, \ldots, m .$$

In view of (1.2),
$$(f, f) \geq 0 .$$
So (f, g) is, generally, a quasiscalar product in $\mathfrak{L}$.

Define the operator A_0 on $\mathfrak{L}$ as
$$\mathcal{D}(A_0) = \mathfrak{L} ,$$

$$(A_0 f)(\lambda) = \lambda f(\lambda) = \sum_{j=0}^{n} \xi_j \lambda^{j+1} .$$

The operator A_0 is Hermitian. It is not difficult to verify that
$$\Phi(f, z) = f(z)$$

is a universal directing functional in the sense of Definition 2.8.2 for the operator A_0. This follows from the definition and the fact that the condition $f(z) = 0$ equivalent to (2.8.2) is necessary and sufficient for the equation

$$A_0 g - z g = (\lambda - z) g = f , \quad f \in \mathfrak{L} ,$$

to be solvable in $\mathfrak{L}$.

It is clear that the element
$$u = 1$$

may be chosen as a gauge in the isomorphism (1.2.6) because
$$\Phi(u, z) \equiv 1 .$$

By the formula (2.8.9),

$$(f, g) = \int_{-\infty}^{\infty} \Phi(f, \lambda) \overline{\Phi(g, \lambda)} \, d\tau(\lambda) = \int_{-\infty}^{\infty} f(\lambda) \overline{g(\lambda)} \, d\sigma(\lambda) , \qquad (1.5)$$

where, in accordance with (2.8.19),

$$\tau(\lambda) = \int_{0}^{\lambda} d(E_t \hat{u} , \hat{u}) ,$$

E_t, $t \in \mathbb{R}^1$, is an arbitrary spectral function of the operator A associated with A_0 in the manner of subsection 2.8.2,

$$\sigma(\lambda) = \int_{-\infty}^{\lambda} d(E_t \hat{u} , \hat{u}) = (E_\lambda \hat{1}, \hat{1})$$

is obtained from $\tau(\lambda)$ by the appropriate renormalization. Setting

$$f(\lambda) = \lambda^k, \quad g(\lambda) = 1,$$

in (1.4) and (1.5), we get

$$s_k = \int_{-\infty}^{\infty} \lambda^k \, d\sigma(\lambda), \quad k \in \mathbb{N}_0,$$

which is what had to be proved. $\square$

1.2   The operator A generated on $\mathfrak{H}_\mathfrak{L}$ by A_0 is real with respect to the involution

$$Jf(\lambda) = \overline{f(\overline{\lambda})}.$$

Therefore the defect numbers of this operator are equal. As was noted in subsection 2.8.5, they are $(0,0)$ or $(1,1)$.

Definition 1.1 *We shall call the moment problem* (1.1) *indefinite if the deficiency index of the operator A is* (1,1).

In the indefinite case, the representation (1.1) is not unique insofar as the function $\sigma(\lambda)$ appearing in this representation is nonuniquely determined by the sequence $\{s_k\}_{k \in \mathbb{N}_0}$.
   Theorem 2.8.4 shows that in order for the power moment problem (1.1) to be indefinite, it is necessary for the form (1.2) to be strongly positive. The latter means that the equality sign in (1.2) is possible if and only if $\xi_0 = \xi_1 = \cdots = \xi_n = 0$, or

$$(f, f) > 0 \quad \text{if } f \neq 0, \tag{1.6}$$

which amounts to the same.
   In what follows we suppose that the condition (1.6) is fulfilled. Then A is the closure of A_0 in the Hilbert space $\mathfrak{H} = \overline{\mathfrak{L}}$. The arguments of subsection 2.8.5 enable us to conclude that in the case of the indefinite moment problem the operator A is entire with respect to the entire gauge $u = 1$ ($M = \mathbb{C}^1$). Moreover, since

$$\lim_{|z| \to \infty} \frac{\log |f(z)|}{|z|} = 0,$$

the type of the operator A, by Theorem 2.5.2, is minimal, i.e.

$$\lim_{|z| \to \infty} \frac{\log \nabla_M(z)}{|z|} = 0, \quad \text{where} \quad \nabla_M(z) = \max_{f \in \mathfrak{H}} \frac{\|f_M(z)\|}{\|f\|}.$$

Consequently we have proved the following theorem.

Theorem 1.2 *The operator A associated with the indefinite moment problem* (1.1) *is an entire operator of minimal type whose deficiency index is* (1,1).

1.3   Since the operator A is entire, the linear mapping

$$f \mapsto f_M(z) = f(z) \cdot 1$$

is, by Theorem 2.2.1, continuous, hence

$$\nabla_M(z) < \infty .$$

Therefore the function $\nabla_M(z)$ may be defined as

$$\nabla_M(z) = \sup_{f \in \mathfrak{L}} \frac{|f(z)|}{\|f\|} . \tag{1.7}$$

It should be noted that this definition is also meaningful in the case where the deficiency index of the operator A is $(0,0)$ if we allow the function $\nabla_M(z)$ to take the value ∞.

Theorem 1.3 *The deficiency index of the operator A is $(1,1)$ if and only if there exists at least one point $z_0 : \Im z_0 \neq 0$ such that*

$$\nabla_M(z_0) < \infty .$$

Proof. For a point $z : \Im z \neq 0$ we put

$$\mathfrak{M}_z^\circ = (A_0 - zI)\mathfrak{L} = (A - zI)\mathfrak{L} .$$

The relation $A = \overline{A_0}$ implies $\mathfrak{M}_z = \overline{\mathfrak{M}_z^\circ}$. Since the condition $g = f - h \cdot 1 \in \mathfrak{M}_z^\circ$ as $f \in \mathfrak{L}$, $h \in \mathbb{C}^1$, is equivalent to the equality $h = f(z)$, the definition (1.7) is equivalent to

$$\nabla_M(z) = \sup_{h \in \mathbb{C}^1, g \in \mathfrak{M}_z^\circ} \frac{|h|}{\|g - h \cdot 1\|} = \sup_{h \in \mathbb{C}^1, g \in \mathfrak{M}_z} \frac{|h|}{\|g - g \cdot 1\|} .$$

So,

$$\nabla_M^{-1}(z) = \sin \alpha ,$$

where α is the angle of inclination of $M = \mathbb{C}^1$ to $\mathfrak{M}_z$. Thus, if $\nabla_M(z_0) < \infty$, $\Im z_0 \neq 0$, then the spaces $M = \mathbb{C}^1$ and $\mathfrak{M}_{z_0}$ are disjoint, hence $\dim \mathfrak{N}_{\bar{z}_0} = \dim \mathfrak{m}_{z_0} \mathfrak{H} = 1$.   □

1.4   Recall that we consider the case where $(f, f) > 0$ as $f \neq 0$. Denote by $\{\mathcal{D}_k(\lambda)\}_{k=0}^{\infty}$ the complete orthonormal system in $\mathfrak{H}$ which is obtained as a result of the successive orthogonalization and normalization of the sequence

$$1, \lambda, \lambda^2, \dots, \lambda^n, \dots . \tag{1.8}$$

The functions $\mathcal{D}_k(\lambda)$ corresponding to the system (1.8) are polynomials of degree k.

It follows from Theorem 2.4.1 that, in the case of the indefinite power moment problem, the estimate

$$\nabla_M^2(z) = \sum_{k=0}^{\infty} |D_k(z)|^2 < \infty \tag{1.9}$$

holds.

The equality

$$\nabla_M^2(z) = \sum_{k=0}^{\infty} |D_k(z)|^2 \tag{1.10}$$

is valid in the general situation too. Indeed, $\mathfrak{M}_z^\circ$, $\Im z \neq 0$, coincides with the set of all the polynomials which vanish at the point z. Therefore

$$\mathfrak{L} = \mathfrak{M}_z^\circ \dotplus \mathbb{C}^1 .$$

Let us find the distance $d(z)$ from the vector 1 to the subspace $\mathfrak{M}_z$. We have

$$d(z) = \inf_{f \in \mathfrak{M}_z} \|1 - f\| = \inf_{f : f(z)=0} \|1 - f\|$$

$$= \inf_{g : g(z)=1} \|g\| = \inf_{g \in \mathfrak{L}} \left\| \frac{g}{g(z)} \right\| = \inf_{n \in \mathbb{N}_0, c_k \in \mathbb{C}^1} \left\| \frac{\sum\limits_{k=0}^{n} c_k D_k}{\sum\limits_{k=0}^{n} c_k D_k(z)} \right\|$$

$$= \inf_{n \in \mathbb{N}_0, c_k \in \mathbb{C}^1} \frac{\sqrt{\sum\limits_{k=0}^{n} |c_k|^2}}{\left| \sum\limits_{k=0}^{n} c_k D_k(z) \right|} = \inf_{n \in \mathbb{N}_0, c_k \in \mathbb{C}^1} \left(\left| \sum_{k=0}^{n} c_k D_k(z) \right| \cdot \left(\sum_{k=0}^{n} |c_k|^2 \right)^{-\frac{1}{2}} \right)^{-1}$$

$$= \inf_{n \in \mathbb{N}_0} \frac{1}{\left(\sum\limits_{k=0}^{n} |D_k(z)|^2 \right)^{1/2}} = \frac{1}{\left(\sum\limits_{k=0}^{\infty} |D_k(z)|^2 \right)^{1/2}} ,$$

and if $1 \in \mathfrak{M}_z$, then $d(z) = 0$, that is

$$\sum_{k=0}^{\infty} |D_k(z)|^2 = \infty = \nabla_M^2(z) .$$

The next statement is known as the Hamburger criterion. It follows from (1.9), (1.10) and Theorem 1.3.

Theorem 1.4 *In order that the power moment problem (1.1) be indefinite, it is necessary and sufficient that*

$$\sum_{k=0}^{\infty} |D_k(z_0)|^2 < \infty$$

for at least one point $z_0 : \Im z_0 \neq 0$.

1.5   By Theorem 2.1.2, the set of all polynomials is dense in the space $L_2(\mathbb{R}^1, d\sigma)$, where $\sigma(\lambda)$ is a distribution function of the operator A associated with the gauge $u = 1$, if and only if $\sigma(\lambda)$ is generated by an orthogonal spectral function of this operator. We call such a function $\sigma(\lambda)$ a canonical solution of the moment problem (1.1). Applying Theorems 2.5.3 and 2.7.3 to the case under consideration, we obtain the following assertion.

Theorem 1.5 *Every canonical solution $\sigma(\lambda)$ of the indefinite power moment problem is a step function. The set of its jump points coincides with the set of all zeroes of the entire function of minimal type*

$$\Phi_\xi(z) = (z - \xi) \sum_{k=0}^{\infty} \mathcal{D}_k(\xi)\mathcal{D}_k(z) \,,$$

where ξ is a certain real number. If a number ξ runs through the whole real axis, then the function $\sigma(\lambda)$ passes the set of all the canonical solutions $\sigma_\xi(\lambda)$ of the moment problem (1.1). Moreover, the inequality

$$\sigma_\xi(\xi + 0) - \sigma_\xi(\xi - 0) \geq \sigma(\xi + 0) - \sigma(\xi - 0)$$

holds for all the solutions $\sigma(\lambda)$ of this problem.

1.6   Following the notation (2.7.4), we introduce the polynomials

$$\mathcal{E}_k(z) = \left(R_z(\mathcal{D}_k - \mathcal{D}_k(z) \cdot 1), 1 \right) = \int_{-\infty}^{\infty} \frac{\mathcal{D}_k(\lambda) - \mathcal{D}_k(z)}{\lambda - z} \, d\sigma(\lambda), \qquad (1.11)$$

where R_z and $\sigma(\lambda)$ are a generalized resolvent and a distribution function associated with the gauge $u = 1$ of the operator A, respectively. It follows from (1.11) that the degree of $\mathcal{E}_k(z)$ is $k - 1$. As has been shown in Section 2.7, these polynomials do not depend on the choice of a generalized resolvent R_z. Set

$$q_1(z) = -z \sum_{k=0}^{\infty} \mathcal{D}_k(z)\mathcal{D}_k(0) \,,$$

$$q_0(z) = 1 - z \sum_{k=0}^{\infty} \mathcal{D}_k(z)\mathcal{E}_k(0) \,,$$

$$\qquad\qquad (1.12)$$

$$p_1(z) = 1 + z \sum_{k=0}^{\infty} \mathcal{E}_k(z)\mathcal{D}_k(0) \,,$$

$$p_0(z) = z \sum_{k=0}^{\infty} \mathcal{E}_k(z)\mathcal{E}_k(0).$$

In the case of the indefinite moment problem the series in (1.12) converge uniformly in every bounded domain of the complex plane.

Applying Theorem 2.7.2 to the indefinite problem (1.1), we arrive at the statement known as the Nevanlinna theorem, which gives an effective description of all solutions of the power moment problem.

Theorem 1.6 *The Nevanlinna matrix*

$$U(z) = \begin{pmatrix} p_1(z) & p_0(z) \\ q_1(z) & q_0(z) \end{pmatrix}$$

whose entries are given by (1.12), is assigned to the indefinite power moment problem (1.1) so that the formula

$$\int_{-\infty}^{\infty} \frac{d\sigma(\lambda)}{\lambda - z} = \frac{p_1(z)\,\tau(z) + p_0(z)}{q_1(z)\,\tau(z) + q_0(z)}$$

determines a one-to-one correspondence between the set of all solutions $\sigma(\lambda)$ of this problem and the set of all R-functions $\tau(z)$. The function $\sigma(\lambda)$ is a canonical solution of the problem if and only if $\tau(z) \equiv t$, where t is a real constant including $t = \infty$.

2   Continuation Problem for Positive Definite Functions

2.1   We start with the definition.

Definition 2.1 *The function $F(t)$, $t \in (-a, a)$, $0 < a \le \infty$, is called positive definite on the interval $(-a, a)$ if*

$$\sum_{j,k=1}^{n} F(t_j - t_k)\xi_j \bar{\xi}_k \ge 0, \quad n \in \mathbb{N}, \quad t_j \in [0, a), \tag{2.1}$$

for any $\xi_j \in \mathbb{C}^1$, $j = 1, \ldots, n$.

It follows immediately from (2.1) that a positive definite function $F(t)$ given on $(-a, a)$ possesses the Hermitian symmetry property

$$F(-t) = \overline{F(t)}.$$

Putting

$$n = 2, \quad t_1 = 0, \quad t_2 = t,$$

in (2.1), we obtain

$$F(0)\,|\xi_1|^2 + 2\,\Re\left(F(t)\,\bar{\xi}_1\xi_2\right) + F(0)\,|\xi_2|^2 \ge 0,$$

whence

$$\left|F(t)\right| \le F(0), \quad t \in (-a, a) .$$

If we set there

$$n = 3, \quad t_1 = 0, \quad t_2 = t, \quad t_3 = s, \quad \xi_1 = \xi, \, \xi_2 = \eta, \quad \xi_3 = -\eta ,$$

we find that

$$F(0) \left|\xi\right|^2 + 2 \Re\left((F(t) - F(s))\, \bar{\xi}\eta\right) + 2\left(F(0) - \Re F(t - s)\right) |\eta|^2 \ge 0,$$

which implies

$$\left|F(t) - F(s)\right| \le 2\left(F(0) - \Re F(t - s)\right) F(0) .$$

The last inequality shows that if $\Re F(t)$ is continuous at the point $t = 0$, then the function $F(t)$ is uniformly continuous on each closed interval $[c, d] \subset (-a, a)$. So, if $a < \infty$ and $\Re F(t)$ is continuous at 0, then setting

$$F(\pm a) = \lim_{t \to \pm a} F(t) ,$$

we may always assume that $F(t)$ is continuous on the closed interval $[-a, a]$.

Denote by $\mathfrak{P}_a$, $0 < a \le \infty$, the class of all continuous functions positive definite on $(-a, a)$. The following theorem gives a criterion for a function to fall into the class $\mathfrak{P}_a$.

Theorem 2.1 *In order that the function $F(t)$ belong to $\mathfrak{P}_a$, it is necessary and sufficient that it admit a representation of the form*

$$F(t) = \int_{-\infty}^{\infty} e^{it\lambda}\, d\sigma(\lambda), \quad t \in (-a, a) , \tag{2.2}$$

where $\sigma(\lambda)$ is a distribution function, i.e. a non-decreasing bounded function normalized by the conditions

$$\sigma(-\infty) = 0, \, \sigma(\lambda - 0) = \sigma(\lambda) .$$

Proof. It is easily checked that any function of the form (2.2) is continuous and the inequality (2.1) is fulfilled for it.

Let us prove the converse. To this end denote by $\mathfrak{L}$ the linear span of functions continuous on the interval $[0, a)$ which vanish in some left neighbourhood (depending on a function) of the point a, and define a bilinear functional on $\mathfrak{L}$ as

$$(f, g) = \int_0^a \int_0^a F(s - t) f(s) \overline{g(t)}\, ds\, dt , \quad f, g \in \mathfrak{L} . \tag{2.3}$$

In view of (2.1), the functional (2.3) has all the properties of a quasiscalar product.

Let $\mathcal{D}(A_0)$ be the set of all continuously differentiable functions $f(t) \in \mathfrak{L}$ satisfying the condition

$$f(0) = 0 . \tag{2.4}$$

On $\mathcal{D}(A_0)$ we define the operator A_0 as follows:

$$A_0 f = if' , \quad f \in \mathcal{D}(A_0) .$$

The operator A_0 is Hermitian. Indeed, the set $\mathcal{D}(A_0)$ is quasidense in $\mathfrak{L}$. This results from the boundedness of $F(t)$ and the fact that every function $f(t) \in \mathfrak{L}$ can be approximated in the $L_2(0,a)$-norm by functions $f_n(t) \in \mathcal{D}(A_0)$, whence

$$(f - f_n, f - f_n) = \int_0^a \int_0^a F(s-t)(f(s) - f_n(s))\overline{(f(t) - f_n(t))} \, ds \, dt \to 0$$

as $n \to \infty$. Next, due to (2.4),

$$(A_0 f, g) = i \int_0^a \int_0^a F(s-t)f'(s)\overline{g(t)} \, ds \, dt = i \int_0^a f'(s) \int_{s-a}^s F(\tau)\overline{g(s-\tau)} \, d\tau \, ds$$

$$= i \int_{-a}^0 F(\tau) \int_0^{\tau+a} f'(s)\overline{g(s-\tau)} \, ds \, d\tau + i \int_0^a F(\tau) \int_\tau^a f'(s)\overline{g(s-\tau)} \, ds \, d\tau$$

$$= i \int_{-a}^0 F(\tau) \left(\overline{g(s-\tau)}f(s) \Big|_{s=0}^{s=\tau+a} - \int_0^{\tau+a} f(s)\,\overline{g'(s-\tau)} \, ds \right) d\tau$$

$$+ i \int_0^a F(\tau) \left(\overline{g(s-\tau)}f(s) \Big|_{s=\tau}^{s=a} - \int_\tau^a f(s)\,\overline{g'(s-\tau)} \, ds \right) d\tau$$

$$= -i \int_{-a}^0 F(\tau) \int_0^{\tau+a} f(s)\,\overline{g'(s-\tau)} \, ds \, d\tau - i \int_0^a F(\tau) \int_\tau^a f(s)\,\overline{g'(s-\tau)} \, ds \, d\tau$$

$$= -i \int_0^a f(s) \int_0^s F(\tau)\,\overline{g'(s-\tau)} \, d\tau \, ds - i \int_0^a f(s) \int_{s-a}^0 F(\tau)\,\overline{g'(s-\tau)} \, d\tau \, ds$$

$$= i \int_0^a f(s) \int_s^0 F(s-t)\,\overline{g'(t)} \, dt \, ds + i \int_0^a f(s) \int_a^s F(s-t)\,\overline{g'(t)} \, dt \, ds$$

$$= \int_0^a \int_0^a F(s-t)f(s)\,\overline{ig'(t)} \, ds \, dt = (f, A_0 g)$$

as $f, g \in \mathcal{D}(A_0)$.

It is not difficult to make sure that

$$\Phi(f, z) = \int_0^a e^{izt} f(t)\, dt\,, \quad f \in \mathfrak{L}\,, \tag{2.5}$$

is a universal directing functional for the operator A_0. In fact, if the equation

$$A_0 g - zg = f, \quad f \in \mathfrak{L}\,, \tag{2.6}$$

has a solution, then this solution is unique and can be written in the form

$$g(t) = -i \int_0^t e^{-iz(t-s)} f(s)\, ds\,.$$

But for the function $g(t)$ to belong to $\mathcal{D}(A_0)$, it is necessary and sufficient that

$$\int_0^a e^{-iz(t-s)} f(s)\, ds = 0\,.$$

Thus, the equation (2.6) is solvable if and only if

$$\Phi(f, z) = \int_0^a e^{izs} f(s)\, ds = 0\,.$$

By Theorem 2.8.1

$$(f, g) = \int_0^a \int_0^a F(s - t) f(s) \overline{g(t)}\, ds\, dt = \int_{-\infty}^{\infty} \Phi(f, \lambda) \overline{\Phi(g, \lambda)}\, d\tau(\lambda)$$

$$= \int_{-\infty}^{\infty} \int_0^a e^{i\lambda s} f(s)\, ds \int_0^a e^{-i\lambda t}\, \overline{g(t)}\, dt\, d\tau(\lambda), \quad f, g \in \mathfrak{L}\,, \tag{2.7}$$

where

$$\tau(\lambda): \tau(0) = 0, \quad \tau(\lambda - 0) = \tau(\lambda)$$

is a non-decreasing function. It remains only to notice that

$$\varphi_\varepsilon^{t_0}(t) = \frac{1}{\varepsilon} \varphi\left(\frac{t - t_0}{\varepsilon}\right), \quad t_0 \in [0, a)\,,$$

is a Cauchy sequence with respect to the quasiscalar product (2.3); here $\varphi(t) \in \mathfrak{L}$ is chosen so that

$$\operatorname{supp}\varphi \subset [t_0, t_0 + \delta), \quad \int_{t_0}^{t_0+\delta} \varphi(t)\, dt = 1\,,$$

δ is a fixed sufficiently small number. Put

$$\delta_{t_0}(t) = \lim_{\varepsilon \to 0} \varphi_\varepsilon^{t_0}(t) \in \overline{\mathfrak{L}} .$$

Substituting $f = \varphi_\varepsilon^{t_0}(t)$, $g = \varphi_\varepsilon^{s_0}(t)$ into (2.7) and passing to the limit as $\varepsilon \to 0$, we get

$$F(s_0 - t_0) = \int_{-\infty}^{\infty} e^{i\lambda(s_0 - t_0)} \, d\tau(\lambda), \quad s_0, t_0 \in [0, a) .$$

Hence, $F(t)$ is represented in the form

$$F(t) = \int_{-\infty}^{\infty} e^{i\lambda t} \, d\tau(\lambda), \quad t \in (-a, a) . \tag{2.8}$$

It follows among other things that the non-decreasing function $\tau(\lambda)$, $\lambda \in \mathbb{R}^1$ is bounded and

$$\int_{-\infty}^{\infty} d\tau(\lambda) = F(0) .$$

The function $\sigma(\lambda) : \sigma(-\infty) = 0$ in (2.2) is the result of the corresponding renormalization of the function $\tau(\lambda)$ in (2.8). $\qquad \square$

In what follows we may suppose without loss of generality that $F(0) = 1$. It should also be noted that in the case of a finite a the representation (2.2) holds true on the closed interval $[-a, a]$ because the function $F(t)$ is continuous there.

2.2   The operator A associated with A_0 in accordance with subsection 2.8.2 is real under the involution

$$Jf(t) = \overline{f(-t)} .$$

Therefore its defect numbers are equal. Since the operator A_0 has a universal directing functional, the deficiency index of A is $(0,0)$ or $(1,1)$. In the latter, $\mathfrak{H}_{\mathfrak{L}} = \overline{\mathfrak{L}}$ (the closure is taken in the norm given by the scalar product (2.3)) is a Hilbert space (denoted by $\mathfrak{H}$), and the operator A is entire with respect to the gauge $u = \delta_0(t)$ ($\Phi(z, \delta_0) \equiv 1$).

If $a = \infty$, then the function $\sigma(\lambda)$ in the representation (2.2) is uniquely determined by the known inversion formula for the Fourier transform. The case of $a < \infty$ differs in principle from the previous one. If the deficiency index of A is $(0,0)$, then the measure $d\sigma(\lambda)$ in the representation (2.2) is unique. But this is not the case when the deficiency index of A is $(1,1)$. The general form of the distribution function $\sigma(\lambda)$ in (2.2) is given, according to (2.8.19), by the formula

$$\sigma(\lambda) = (E_\lambda u, u), \quad u = \delta_0(t) , \tag{2.9}$$

where E_λ is a spectral function of the operator A.

Theorem 2.1 gives in essence an answer to the very intriguing question: is it possible to continue a function positive definite on a finite interval $[-a, a]$ onto the whole real axis having retained the positive definiteness property? The answer is "Yes". Indeed, such a function coincides on $[-a, a]$ with the positive definite on $\mathbb{R}^1$ function

$$\int_{-\infty}^{\infty} e^{i\lambda t} \, d\sigma(\lambda), \quad t \in \mathbb{R}^1 ,$$

generated by the representation (2.2). So the desired continuation is constructed. Insofar as in the case of a finite interval the measure $d\sigma(\lambda)$ in (2.9) is, generally, nonunique, the continuation of the positive definite function $F(t)$, $t \in [-a, a]$, $a < \infty$, may also be nonunique. Depending on whether the continuation is unique or not, two cases are distinguishable in the continuation problem for a positive definite function, namely: definite if the continuation is unique, and indefinite in the opposite case. In the latter case, which we shall mainly be interested in, the description of all possible positive definite continuations of $F(t)$, $t \in [-a, a]$, $a < \infty$, from the interval $[-a, a]$ to $\mathbb{R}^1$ is of great importance. Such a description will be given later.

Theorem 2.2 *In the indefinite case, the form* (2.3) *determines a scalar product in* $\mathfrak{L}$, *the deficiency index of the operator A is* (1,1). *This operator is entire of normal type, and the indicator $h_{\nabla_M}(\varphi)$ of the function $\nabla_M(z)$ is given by the formula*

$$h_{\nabla_M}(\varphi) = \varlimsup_{r \to \infty} \frac{\log \nabla_M(re^{i\varphi})}{r} = \frac{a}{2} \left(|\sin \varphi| - \sin \varphi \right), \quad 0 \leq \varphi \leq 2\pi . \quad (2.10)$$

Proof. Only the formula (2.10) need be proved because the rest was established above. Taking the element

$$u = \delta_0(t)$$

as a gauge, we obtain

$$f_M(z) = \Phi(f, z)u = \int_0^a e^{izt} f(t) \, dt \, u \quad (2.11)$$

for any $f \in \mathfrak{L}$. By Theorem 2.3.2 all the vector-functions $f_M(z)$, $f \in \mathfrak{H}$, are entire. The equality (2.10) for the indicator $h_{\nabla_M}(\varphi)$ of the function

$$\nabla_M(z) = \sup_{f \in \mathfrak{L}} \frac{\|f_M(z)\|}{\|f\|} = \sup_{f \in \mathfrak{L}} \frac{|f_u(z)|}{\|f\|} \quad (2.12)$$

means that the indicator diagram of this function is the interval $[-ia, 0]$ (see subsection 1.4.5). It follows from (2.11) that

$$h_{f_u}(\varphi) = \varlimsup_{r \to \infty} \frac{\log \|f_M(re^{i\varphi})\|}{r} \leq \frac{a}{2} \left(|\sin \varphi| - \sin \varphi \right), \quad 0 \leq \varphi \leq 2\pi .$$

This inequality shows that the indicator diagram of every function $f_u(z)$, $f \in \mathfrak{L}$, is contained in the interval $[-ia, 0]$ of the imaginary axis.

On the other hand, any point $-is$, $0 \leq s \leq a$, is the indicator diagram of the function e^{izs}, which is the image of the element $\delta_s(t)$ under the isomorphism

$$f \mapsto f_u(z) = \Phi(f, z) \,.$$

By Theorem 2.5.2 the segment $[-ia, 0]$ is the indicator diagram of the function $\nabla_M(z)$.   □

2.3 In this subsection some examples are given of both definite and indefinite cases in the continuation problem for positive definite functions.

a)   Let $F(t)$, $t \in \mathbb{R}^1$, be a continuous non-negative even convex function such that $F(t) \to 0$ as $|t| \to \infty$. Such a function is positive definite. Indeed, as is known, the integral

$$s(\lambda) = \frac{2}{\pi} \int\limits_0^\infty F(t) \cos \lambda t \, dt \quad \text{converges for } \lambda \neq 0, \quad s(\lambda) \geq 0, \quad \int\limits_{-\infty}^\infty s(\lambda) \, d\lambda < \infty \,,$$

and the representation

$$F(t) = \int\limits_{-\infty}^\infty e^{i\lambda t} s(\lambda) \, d\lambda = \int\limits_{-\infty}^\infty e^{i\lambda t} \, d \int\limits_{-\infty}^\lambda s(\mu) \, d\mu \,, \quad t \in \mathbb{R}^1 \,,$$

holds, whence the positive definiteness of $F(t)$ follows.

If now $F(t)$, $t \in (-a, a)$, $0 < a < \infty$, is a continuous even positive convex function, then it can be continued to the whole real axis having preserved all these properties and so that $\lim\limits_{|t| \to \infty} F(t) = 0$. Since the continuation is positive definite, so is the original function. Since the number of possible continuations is infinite, the continuation problem for the given function $F(t)$ ($a < \infty$) is indefinite.

b)   As has been observed by P. Levy, a function $F(t) \in \mathfrak{P}_\infty$ analytic in the disk with center $z = 0$ and radius r is analytic in the strip $|\Im z| \leq r$. It follows that the continuation problem for a function $F(t) \in \mathfrak{P}_a$ analytic in some neighbourhood of the point $z = 0$ is always definite.

c)   Put

$$F(t) = 1 - |t|, \quad t \in [-2, 2] \,.$$

The Fourier-series expansion of $F(t)$

$$F(t) = \frac{8}{\pi^2} \sum_{k=1}^\infty \frac{\cos \dfrac{2k-1}{2} \pi t}{(2k-1)^2} \quad \text{may be written as} \quad F(t) = \int\limits_{-\infty}^\infty e^{i\lambda t} \, d\sigma(\lambda) \,,$$

where $\sigma(\lambda)$ is a step function whose jump points are $\lambda_k = \pm\dfrac{2k-1}{2}\pi$, the corresponding jumps equal $\dfrac{4}{\pi^2(2k-1)^2}$. So the function $F(t)$ is positive definite. Since

$$\begin{vmatrix} F(0) & F(2) \\ F(2) & F(0) \end{vmatrix} = 0,$$

the form

$$\sum_{i,k=1}^{2} F(t_i - t_k)\xi_i\bar{\xi}_k$$

degenerates. Hence, there exists a vector $f_0 \neq 0 : (f_0, f_0) = 0$. This implies that the measure $d\sigma(\lambda)$ is uniquely determined by $F(t)$. Therefore the continuation problem for this function is definite.

d)   Let a function $F(t)$ be twice continuously differentiable on the segment $[0, a]$, $a < \infty$, and $F'(0) < 0$. Define $F(t)$ as $F(t) = F(-t)$ on the interval $[-a, 0]$. If a is sufficiently small, then the function $F(t)$ is positive definite on $(-a, a)$, and the continuation problem for it from the interval $(-a, a)$ to the whole real axis is indefinite.

Indeed, setting

$$g(t) = \int_0^t f(\xi)\, d\xi, \quad f(t) \in \mathfrak{L},$$

we obtain by integrating by parts $(l < a)$

$$\int_0^l \int_0^l F(t - s)f(t)\overline{f(s)}\, ds\, dt$$

$$= \int_0^l f(t)\left(\int_0^t F(t - s)\overline{f(s)}\, ds + \int_t^l F(t - s)\overline{f(s)}\, ds\right) dt$$

$$= \int_0^l f(t)\left(\overline{g(l)}F(l - t) + \int_0^t F'(t - s)\overline{g(s)}\, ds - \int_t^l F'(s - t)\overline{g(s)}\, ds\right) dt$$

$$= \overline{g(l)} \int_0^l F(l - t)f(t)\, dt$$

$$+ \int_0^l f(t)\left(\int_0^t F'(t - s)\overline{g(s)}\, ds\right) dt - \int_0^l f(t)\left(\int_t^l F'(s - t)\overline{g(s)}\, ds\right) dt$$

$$= F(0)|g(l)|^2 - 2\,F'(0)\int_0^l |g(t)|^2\,dt + 2\,\Re\left(g(l)\int_0^l F'(l-s)g(s)\,ds\right)$$

$$- \int_0^l\int_0^l F''(t-s)g(t)\overline{g(s)}\,dt\,ds \geq F(0)|g(l)|^2 + \gamma\int_0^l |g(s)|^2\,ds$$

$$- 2c\,l^{1/2}|g(l)|\left(\int_0^l |g(s)|^2\,ds\right)^{1/2} - c_1 l\int_0^l |g(s)|^2\,ds$$

$$\geq F(0)|g(l)|^2 + \gamma\int_0^l |g(s)|^2\,ds$$

$$- cl^{1/2}\left(|g(l)|^2 + \int_0^l |g(s)|^2\,ds\right) - c_1 l\int_0^l |g(s)|^2\,ds$$

$$= \left(F(0) - cl^{1/2}\right)|g(l)|^2 + \left(\gamma - cl^{1/2} - c_1 l\right)\int_0^l |g(s)|^2\,ds\,,$$

where
$$\gamma = -F'(0) > 0,\quad c = \sup_{t\in[0,l]}|F'(t)|,\quad c_1 = \sup_{t\in[0,l]} F''(t)\,.$$

If we choose l so small that $F(0) - c\sqrt{l} \geq 0$ and $\gamma - c\sqrt{l} - c_1 l \geq 0$, then

$$\int_0^l\int_0^l F(t-s)f(t)\overline{f(s)}\,dt\,ds \geq 0\,,$$

hence the function $F(t)$ is positive definite on $(-l,l)$.

Now let $F_1(t)$ be a twice differentiable positive definite function which admits a nonunique continuation from an arbitrary interval $(-a,a)$ to the whole real axis (for instance, the function appearing in the example a)). Then the function $F_2(t) = F(t) - \varepsilon F_1(t)$ (ε is small enough) is positive definite in a sufficiently small interval $(-l,l)$. Therefore, the continuation problem for the function $F(t) = F_2(t) + \varepsilon F_1(t)$ from any interval $(-l,l)$ to $\mathbb{R}^1$ is indefinite.

e)   Whatever $a < \infty$, a function of the form (2.2) is nonuniquely continued to a positive definite function on the whole real axis if

$$\int_{-\infty}^{\infty} \frac{\log \sigma'(\lambda)}{1+\lambda^2}\,d\lambda > -\infty\,, \tag{2.13}$$

where $\sigma'(\lambda)$ is understood to be the derivative of the absolutely continuous part of $\sigma(\lambda)$.

To verify this we need the following result.

Theorem 2.3 *The relation*

$$\overline{\{\Phi(f,\lambda)\}}_{f\in\mathfrak{L}} \;=\; \overline{\{e^{i\lambda t}\}}_{t\in[-a,a]}$$

is valid, where the closure is taken in the $L_2(\mathbb{R}^1, d\sigma)$*-norm.*

Proof. According to (2.5)

$$\Phi(\delta_t,\lambda) \;=\; \lim_{\varepsilon\to 0} \frac{1}{\varepsilon} \int_0^a e^{i\lambda s}\varphi_\varepsilon^t(s)\,ds \;=\; e^{i\lambda t}\,,$$

hence

$$\overline{\{e^{i\lambda t}\}}_{t\in[-a,a]} \;\subseteq\; \overline{\{\Phi(f,\lambda)\}}_{f\in\mathfrak{L}}\,.$$

Conversely, in view of

$$\left(e^{it\lambda}, e^{it\lambda}\right)_{L_2(\mathbb{R}^1, d\sigma)} \;=\; \int_{-\infty}^{\infty} d\sigma(\lambda) \;=\; 1$$

and

$$\left\|e^{it\lambda} - e^{is\lambda}\right\|_{L_2(\mathbb{R}^1, d\sigma)} \;=\; 2 - 2\,\Re \int_{-\infty}^{\infty} e^{i(t-s)\lambda}\,d\sigma(\lambda) \;=\; 2 - 2\,\Re\,F(t-s)\,,$$

we have

$$\left\|e^{it\lambda} - e^{is\lambda}\right\|_{L_2(\mathbb{R}^1, d\sigma)} \to 0 \quad \text{as} \quad (t-s)\to 0\,,$$

i.e. $e^{it\lambda}$ is a uniformly continuous vector-function on $[-a,a]$. Therefore for each $f\in\mathfrak{L}$ the limit in the $L_2(\mathbb{R}^1, d\sigma)$-norm of the integral sums

$$\sum e^{it_j\lambda}\, f(t_j)\Delta t_j$$

exists when $\max \Delta t_j \to 0$. On the other hand,

$$\int_0^a e^{it\lambda}\, f(t)\,dt$$

converges for every $\lambda\in\mathbb{R}^1$, $f\in\mathfrak{L}$. So the function

$$\Phi(f,\lambda) \;=\; \int_0^a e^{it\lambda}\, f(t)\,dt$$

is a limit in the $L_2(\mathbb{R}^1, d\sigma)$-norm of certain linear combinations of $e^{it_j\lambda}$. Thus,

$$\overline{\{\Phi(f,\lambda)\}}_{f\in\mathfrak{L}} \;\subseteq\; \overline{\{e^{i\lambda t}\}}_{t\in[-a,a]}\,. \qquad \square$$

But the condition

$$\int\limits_{-\infty}^{\infty} \frac{|\log \sigma'(\lambda)|}{1 + \lambda^2} \, d\lambda = \infty$$

(see, for instance, Akhiezer [1,2])

is necessary and sufficient for the set $\{e^{i\lambda t}\}_{t\in[0,\infty)}$ to be dense in $L_2(\mathbb{R}^1, d\sigma)$. Consequently, if the inequality (2.13) holds, then

$$\overline{\{e^{i\lambda t}\}}_{t\in[0,2a]} \neq L_2(\mathbb{R}^1, d\sigma) \, .$$

By Theorems 2.3 and 2.1.2 $\sigma(\lambda)$ is not a canonical distribution function of the operator A, hence $\sigma(\lambda)$ is determined by the formula (2.9) where the spectral function E_λ is not orthogonal. In the example under consideration the continuation problem cannot be definite.

The condition (2.13) makes it possible to construct a number of examples of positive definite functions which are nonuniquely continued onto $\mathbb{R}^1$.

f)   Let $R(\bar{z}) = \overline{R(z)}$ be a rational function such that $z = \infty$ is its zero of multiplicity not less than 2. Suppose in addition that $R(z)$ has no poles on the real axis and $R(x) > 0$ when $x \in \mathbb{R}^1$. Then the function

$$F(x) = \int\limits_{-\infty}^{\infty} R(t) \, e^{ixt} \, dt \, , \quad x \in [-a, a], \ a < \infty \, ,$$

can be represented in the form

$$F(x) = \int\limits_{-\infty}^{\infty} e^{ixt} \, d\sigma(t) \, , \quad x \in [-a, a] \, ,$$

where

$$\sigma(t) = \int\limits_{-\infty}^{t} R(x) \, dx \, , \quad t \in \mathbb{R}^1 \, ,$$

hence $F(x) \in \mathfrak{P}_a$. Since

$$\int\limits_{-\infty}^{\infty} \frac{|\log \sigma'(t)|}{1 + t^2} \, dt = \int\limits_{-\infty}^{\infty} \frac{|\log R(t)|}{1 + t^2} \, dt < \infty \, ,$$

the continuation problem for the function $F(x)$ is indefinite.

It should be remarked that the function

$$F(x) = \frac{1}{\pi} \int_{-\infty}^{\infty} \frac{e^{ixt}}{1+t^2}\, dt = e^{-|x|}, \quad x \in [-a, a], \quad a < \infty,$$

is a concrete realization of the example just considered. It follows from this that $e^{-|x|}$ admits a nonunique continuation onto $\mathbb{R}^1$. Starting from $e^{-\alpha|x|}$, $\alpha > 0$, it is possible to construct positive definite functions which are neither convex nor monotone. For instance, such are the functions of the form

$$F(x) = e^{-\alpha|x|} \sum_{j=1}^{k} \rho_j \cos \beta_j x, \quad \alpha > 0, \ \rho_j > 0, \quad j = 1, 2, \dots, k\,.$$

g)   Let

$$F(t) = \int_{-\infty}^{\infty} e^{i\lambda t}\, d\sigma(\lambda) \in \mathfrak{P}_{\infty}\,. \tag{2.14}$$

If the measure $d\sigma(t)$ is such that the power moments

$$\int_{-\infty}^{\infty} \lambda^k\, d\sigma(\lambda), \quad k \in \mathbb{N}_0\,, \tag{2.15}$$

exist, then the function $F(t)$ is infinitely differentiable at the point $t = 0$ and

$$(-i)^k F^{(k)}(0) = \int_{-\infty}^{\infty} \lambda^k\, d\sigma(\lambda), \quad k \in \mathbb{N}_0\,. \tag{2.16}$$

Conversely, if the function $F(t)$ is $2n$-differentiable at the point $t = 0$, then

$$(\Delta_h^{2n} F)(0) \to F^{(2n)}(0) \quad \text{as} \ h \to 0\,,$$

where

$$(\Delta_h F)(t) = \frac{1}{2h}(F(t+h) - F(t-h))\,.$$

But

$$(\Delta_h e^{i\lambda\cdot})(t) = \frac{i}{h}\sin \lambda h\, e^{i\lambda t}\,;$$

$$(\Delta_h^{2n} e^{i\lambda\cdot})(t) = \left(\frac{i}{h}\sin \lambda h\right)^{2n} e^{i\lambda t}\,.$$

Acting by the difference operator Δ_h^{2n} on both sides of (2.14), we find that

$$(-1)^n \left(\Delta_h^{2n} F\right)(0) = \int\limits_{-\infty}^{\infty} \left(\frac{\sin \lambda h}{h}\right)^{2n} d\sigma(\lambda) \geq \int\limits_{-N}^{N} \left(\frac{\sin \lambda h}{\lambda h}\right)^{2n} \lambda^{2n} d\sigma(\lambda), \quad N > 0.$$

Taking into account the existence of the limit as $h \to 0$ in the left-hand side of the last inequality, we conclude that

$$\int\limits_{-\infty}^{\infty} \lambda^k d\sigma(\lambda) < \infty, \quad k = 1, 2, \ldots, 2n.$$

Thus, in order that a positive definite function $F(t)$ be infinitely differentiable at the point $t = 0$, it is necessary and sufficient that all the moments (2.15) exist. It follows from (2.16) that if a function $F(t) \in \mathfrak{P}_a$, $a < \infty$, is infinitely differentiable at $t = 0$, then the definiteness of the continuation problem for it is equivalent to that of the corresponding moment problem (2.16). According to the classical Carleman result (see, for instance, Berezansky [1]) this is the case if

$$\sum\limits_{n=1}^{\infty} \frac{1}{\sqrt[2n]{(-1)^n F^{(2n)}(0)}} = \infty.$$

On the other hand, we obtain another condition sufficient for the power moment problem to be indefinite. Namely, if the sequence $s_0 > 0, s_1, s_2, \ldots$ is represented in the form

$$s_k = \int\limits_{-\infty}^{\infty} t^k d\sigma(t), \tag{2.17}$$

where $\sigma(t)$ is a non-decreasing function of bounded variation such that

$$\int\limits_{-\infty}^{\infty} \frac{|\log \sigma'(t)|}{1 + t^2} dt < \infty,$$

then the moment problem (2.17) is indefinite.

Indeed, in this case,

$$s_k = (-i)^k F^{(k)}(0), \quad k = 0, 1, \ldots,$$

where the function

$$F(t) = \int\limits_{-\infty}^{\infty} e^{i\lambda t} d\sigma(\lambda)$$

is positive definite on an arbitrary finite interval $(-a, a)$.

2.4   Reasoning as in the case of the moment problem, we may define the function $\nabla_M(z)$ (under the assumption that the equality $\nabla_M(z) = \infty$ is possible) by the formula (2.12) irrespective of whether the deficiency index of the operator A is (1,1) or (0,0). It is easily seen that then Theorem 1.3 concerning the power moment problem remains completely true for the case of a positive definite function.

Arrange the elements $\delta_s(t)$ with rational $s \in [0, a]$ and orthogonalize this sequence with respect to the scalar product (2.3) or, equivalently, (2.7). Denote by $\{d_k\}_{k=0}^{\infty}$ the orthonormalized system. Let $\mathcal{D}_k(z)\, u$ $(u = \delta_0(t))$ be the image of d_k under the isomorphism $f \mapsto f_M(z)$. Using Theorem 2.4.1, we arrive at the following assertion.

Theorem 2.4 *In order that the continuation problem for the function (2.2) be indefinite, it is necessary and sufficient that*

$$\sum_{k=0}^{\infty} |\mathcal{D}_k(z)|^2 < \infty$$

for at least one point $z : \Im z \neq 0$.

By Theorems 2.1.2 and 2.3, the set $\{e^{ist}\}$, where s runs through all rational points from $[0, 2a]$, is dense in $L_2(\mathbb{R}^1, d\sigma)$ if and only if the function $\sigma(\lambda)$ in (2.2) is generated by a certain orthogonal spectral function of the operator A. Such a function $\sigma(\lambda)$ is called a canonical solution of the continuation problem for the positive definite function $F(t)$, $t \in [-a, a]$. In complete analogy with the power moment problem the following theorem is established.

Theorem 2.5 *In the indefinite case, every canonical solution $\sigma_\xi(\lambda)$ of the continuation problem for the function $F(t) \in \mathfrak{P}_a$, $a < \infty$, to a function from $\mathfrak{P}_\infty$ is a step function whose jump points coincide with zeroes of the entire function of exponential type a*

$$\Phi_\xi(z) = (z - \xi) \sum_{k=0}^{\infty} \mathcal{D}_k(\xi)\, \mathcal{D}_k(z)\,,$$

where ξ is a certain real number. While ξ passes through the whole real axis, the function $\sigma_\xi(\lambda)$ runs through the set of all canonical solutions of the continuation problem under consideration. Moreover, if $\sigma(\lambda)$ is an arbitrary solution of the continuation problem, then

$$\sigma_\xi(\xi + 0) - \sigma_\xi(\xi) \geq \sigma(\xi + 0) - \sigma(\xi)\,.$$

2.5   Set

$$\mathcal{E}_k(z) = \big(R_z(d_k - \mathcal{D}_k(z)u), u\big) = \int_{-\infty}^{\infty} \frac{\mathcal{D}_k(\lambda) - \mathcal{D}_k(z)}{\lambda - z}\, d\sigma(\lambda)\,, \quad u = \delta_0(t)\,,$$

where R_z and $d\sigma(\lambda)$ are a generalized resolvent and the corresponding distribution function associated with the gauge u of the operator A. By Theorem 2.3, $\mathcal{E}_k(z)$ is a finite linear combination of the functions e^{isz} with rational $s \in [0, 2a]$.

Put

$$q_1(z) = -z \sum_{k=0}^{\infty} \mathcal{D}_k(z)\mathcal{D}_k(0) \,,$$

$$q_0(z) = 1 - z \sum_{k=0}^{\infty} \mathcal{D}_k(z)\mathcal{E}_k(0) \,,$$

$$(2.18)$$

$$p_1(z) = 1 + z \sum_{k=0}^{\infty} \mathcal{E}_k(z)\mathcal{D}_k(0) \,,$$

$$p_0(z) = z \sum_{k=0}^{\infty} \mathcal{E}_k(z)\mathcal{E}_k(0) \,.$$

In the indefinite case these functions are entire of exponential type. Due to Theorem 2.7.2, we obtain an effective description of all solutions of the continuation problem for a positive definite function. The next statement solves this problem.

Theorem 2.6 *The Nevanlinna matrix*

$$U(z) = \begin{pmatrix} p_1(z) & p_0(z) \\ q_1(z) & q_0(z) \end{pmatrix}$$

whose entries are expressed by (2.18) is associated with the indefinite continuation problem for a function $F(t) \in \mathfrak{P}_a$, $a < \infty$, so that the relation

$$\int_{-\infty}^{\infty} \frac{d\sigma(\lambda)}{\lambda - z} = \frac{p_1(z)\,\tau(z) + p_0(z)}{q_1(z)\,\tau(z) + q_0(z)}$$

determines a one-to-one correspondence between the set of all solutions $\sigma(\lambda)$ of this problem and the set of all R-functions $\tau(z)$. The function $\sigma(\lambda)$ gives a canonical solution of the problem if and only if $\tau(z) \equiv t$, where t is a real constant including $t = \infty$.

3   Unitary and Spiral Curves

3.1   Let x_t, $t \in [a, b]$, be a vector-function taking its values in a Hilbert space $\mathfrak{H}$.
The interval $[a, b]$ will be assumed to be closed. If the end a or b is infinite, we do not add it to this interval.

Definition 3.1 *We shall call a continuous vector-function* x_t, $t \in [a, b]$, *a spiral arc if for any* r, $r + s$, $r + t$ *from* (a, b) *the scalar product*

$$(x_{r+s} - x_r, x_{r+t} - x_r)$$

does not depend on r. *If* $a = -\infty$, $b = \infty$, *then the spiral arc is called a spiral line or, briefly, a spiral.*

Thus, for a spiral arc the function

$$B(s, t) = (x_{r+s} - x_r, x_{r+t} - x_r) \tag{3.1}$$

is completely determined by its values at the points $\{s, t\} \in [-T, T] \times [-T, T]$, where $T = b - a$.

Definition 3.2 *The function* $B(s, t)$, $s, t \in [-T, T]$ *defined by (3.1), is called the metric function of the spiral arc* x_t.

If the space $\mathfrak{H}$ is real, then the definition of a spiral arc may be simplified, because the independence on r of the expression

$$(x_{r+s} - x_r, x_{r+t} - x_r) = \frac{1}{2}\left(\|x_{r+t} - x_r\|^2 + \|x_{r+s} - x_r\|^2 - \|x_{r+t} - x_{r+s}\|^2 \right)$$

is equivalent to that of the distance

$$\|x_{r+t} - x_r\|, \quad r, \, r + t \in [a, b]. \tag{3.2}$$

Therefore, in a real space $\mathfrak{H}$ it is possible to define a spiral arc as a continuous curve $x = x_t$, $t \in [a, b]$, such that the function (3.2) does not depend on r.
A straight line and a circle in $\mathbb{R}^2$, a straight line, a circle and a spiral in $\mathbb{R}^3$ are spirals in the sense of the above definition.

Definition 3.3 *A spiral* $\tilde{x}_t$, $t \in \mathbb{R}^1$, *is called a complete continuation of a spiral arc* x_t, $t \in [a, b]$, *if* $\tilde{x}_t = x_t$ *as* $t \in [a, b]$.

Definition 3.4 *We call a continuous vector-function* x_t, $t \in [a, b]$, *in* $\mathfrak{H}$ *a unitary arc if the scalar product* (x_t, x_s) *depends only on the difference* $t - s$, *that is,*

$$(x_t, x_s) = F(t - s). \tag{3.3}$$

If $a = -\infty$, $b = \infty$, then the unitary arc x_t is called a unitary line.

It follows from this definition that every unitary line is bounded. It lies on the sphere $\|x_t\|^2 = F(0)$. Note also that in the case of a finite interval $[a, b]$ we may always suppose $a = -\frac{T}{2}, b = \frac{T}{2}$.

Definition 3.5 *The function $F(t)$, $t \in [-T, T]$ defined by (3.3), is called the correlation function of the unitary arc x_t.*

Theorem 3.1 *A continuous function $F(t)$, $t \in [-T, T]$, is the correlation function of a certain unitary arc x_t, $t \in [-\frac{T}{2}, \frac{T}{2}]$, if and only if it is positive definite.*

Proof. Let $F(t)$, $t \in [-T, T]$, be the correlation function of a unitary arc x_t, $t \in [-\frac{T}{2}, \frac{T}{2}]$. Then

$$\sum_{j,k=1}^{n} F(t_j - t_k)\xi_j\bar{\xi}_k = \left\| \sum_{j=1}^{n} \xi_j x_{t_j} \right\|^2 \geq 0,$$

$$t_j \in [-\frac{T}{2}, \frac{T}{2}], \quad \xi_j \in \mathbb{C}^1, \quad j = 1, 2, \ldots, n, \quad n \in \mathbb{N}.$$

So $F(t) \in \mathfrak{P}_T$.

Conversely, if $F(t) \in \mathfrak{P}_T$, then the function $F(t)$ can be represented in the form

$$F(t) = \int_{-\infty}^{\infty} e^{i\lambda t}\, d\sigma(\lambda), \quad t \in [-T, T], \tag{3.4}$$

where $d\sigma(\lambda)$ is a non-negative finite measure. It is obvious that the arc $x_t = e^{i\lambda t}$ is unitary in the space $L_2(\mathbb{R}^1, d\sigma)$. Its correlation function coincides with $F(t)$. $\square$

It is easily seen that every unitary arc is a spiral arc. The complete unitary continuation of a unitary arc is defined in the same way as a complete spiral continuation of a spiral arc.

Denote by $\mathfrak{L}_x$ the smallest subspace containing a unitary arc x_t, $t \in [-\frac{T}{2}, \frac{T}{2}]$:

$$\mathfrak{L}_x = \text{c.l.s.} \left\{ x_t \right\}_{t \in [-\frac{T}{2}, \frac{T}{2}]}.$$

The following theorem concerning continuation of unitary arcs holds.

Theorem 3.2 *Any unitary arc x_t, $t \in [-\frac{T}{2}, \frac{T}{2}]$, $T < \infty$, in $\mathfrak{H}$ admits a continuation to a unitary line $\tilde{x}_t$, $t \in \mathbb{R}^1$, so that $\tilde{x}_t \in \mathfrak{L}_x$ for all $t \in \mathbb{R}^1$.*

Proof. We may assume without loss of generality $\mathfrak{L}_x = \mathfrak{H}$.

Let $F(t)$ be the correlation function of the arc x_t. In view of Theorems 3.1 and 2.1, the function $F(t)$ is represented in the form (3.4). If this representation is unique, then we take the function $\sigma(\lambda)$. If this is not the case, choose an arbitrary canonical $\sigma(\lambda)$. By Theorems 2.1.2 and 2.3, in both cases

$$\text{c.l.s.} \left\{ e^{i\lambda t} \right\}_{t \in [-\frac{T}{2}, \frac{T}{2}]} = L_2(\mathbb{R}^1, d\sigma).$$

We associate the function $e^{i\lambda t} \in L_2(\mathbb{R}^1, d\sigma)$ with the element $x_t \in \mathcal{L}_x$. The correspondence

$$x_t \mapsto e^{i\lambda t}$$

determines the unitary operator U from $\mathfrak{H}$ onto $L_2(\mathbb{R}^1, d\sigma)$. Setting

$$\widetilde{x}_t = U^{-1} e^{i\lambda t}, \quad t \in \mathbb{R}^1 \,,$$

we obtain the desired complete unitary continuation $\widetilde{x}_t$, $t \in \mathbb{R}^1$, of the arc x_t, $t \in [-\frac{T}{2}, \frac{T}{2}]$.   □

3.2   Let x_t, $t \in (-\infty, 0]$ be a unitary arc in $\mathfrak{H}$ and $F(t)$, $t \in \mathbb{R}^1$, its correlation function. The complete unitary continuations $\widetilde{x}_t$ and $\widetilde{x}'_t$, $t \in \mathbb{R}^1$, of a unitary arc x_t are called *unitarily equivalent* if there exists a unitary operator U acting from $\mathcal{L}_{\widetilde{x}}$ onto $\mathcal{L}_{\widetilde{x}'}$ so that

$$U\widetilde{x}_t = \widetilde{x}'_t \,.$$

Theorem 3.3 *Any two complete unitary continuations of a unitary arc x_t given on the semiaxis $(-\infty, 0]$ are unitary equivalent.*

Proof.   As the correlation function of the unitary arc x_t, $t \in (-\infty, 0]$, is defined on the whole real axis, the map

$$U \sum_{j=1}^n \xi_j \widetilde{x}_{t_j} = \sum_{j=1}^n \xi_j \widetilde{x}'_{t_j}, \quad \xi_j \in \mathbb{C}^1, \quad n \in \mathbb{N},$$

from $\mathcal{L}_{\widetilde{x}}$ onto $\mathcal{L}_{\widetilde{x}'}$ is unitary.   □

Theorem 3.4 *Let a complete unitary continuation $\widetilde{x}_t$, $t \in \mathbb{R}^1$, of a unitary arc x_t, $t \in (-\infty, 0]$, exist. In order that this continuation be unique, it is necessary and sufficient that $\mathcal{L}_x = \mathcal{L}_{\widetilde{x}}$.*

Proof.   Let a unitary arc x_t, $t \in (-\infty, 0]$ have only one complete unitary continuation $\widetilde{x}_t$, $t \in \mathbb{R}^1$. We must prove that $\mathcal{L}_x = \mathcal{L}_{\widetilde{x}}$.

Suppose the contrary, i.e.

$$\mathcal{L}_{\widetilde{x}} = \mathcal{L}_x \oplus \mathcal{L}_x^\perp \,,$$

where $\mathcal{L}_x^\perp$ is the orthogonal complement of $\mathcal{L}_x$ to $\mathcal{L}_{\widetilde{x}}$. Set

$$U = I + U^\perp \,;$$

here $U^\perp$ is an arbitrary unitary operator on $\mathcal{L}_x^\perp$. The line

$$\widetilde{x}'_t = U\widetilde{x}_t, \quad t \in \mathbb{R}^1 \,,$$

is also a complete unitary continuation of x_t, because

$$\left(\widetilde{x}'_t, \widetilde{x}'_s\right) = \left(U\widetilde{x}_t, U\widetilde{x}_s\right) = \left(\widetilde{x}_t, \widetilde{x}_s\right) = F(t - s) \,.$$

If $U^\perp \neq I$, this continuation differs from $\widetilde{x}_t$, contrary to the uniqueness of a complete unitary continuation of x_t.

Conversely, suppose $\mathfrak{L}_x = \mathfrak{L}_{\widetilde{x}}$ but nevertheless there exists a complete unitary continuation $\widetilde{x}_t'$ different from $\widetilde{x}_t$ of the unitary arc x_t, $t \in (-\infty, 0]$. By Theorem 3.3, $\widetilde{x}_t'$ is unitarily equivalent to $\widetilde{x}_t$. So, there exists a unitary operator U such that $\widetilde{x}_t' = U\widetilde{x}_t$. Since this operator leaves fixed the space $\mathfrak{L}_x$, $U = I$ and $\widetilde{x}_t' = \widetilde{x}_t$. □

Now let $d\sigma(\lambda)$ be the measure in the representation (3.4) of the correlation function $F(t)$, $t \in \mathbb{R}^1$, of a unitary arc x_t, $t \in (-\infty, 0]$. Insofar as the function $F(t)$ belongs to $\mathfrak{P}_\infty$, this measure is uniquelly determined by the arc x_t. The map

$$x_t \mapsto e^{i\lambda t}, \quad t \in (-\infty, 0],$$

isometrically transforms $\mathfrak{L}_x$ into $L_2(\mathbb{R}^1, d\sigma)$. We may identify $\mathfrak{L}_x$ with a certain subspace of $L_2(\mathbb{R}^1, d\sigma)$:

$$\mathfrak{L}_x \subset L_2(\mathbb{R}^1, d\sigma).$$

Then the function $\widetilde{x}_t = e^{i\lambda t}$, $t \in \mathbb{R}^1$, is a complete unitary continuation of the arc x_t. Moreover, $\mathfrak{L}_{\widetilde{x}} = L_2(\mathbb{R}^1, d\sigma)$. Therefore, in order that a complete unitary continuation of x_t be unique, it is necessary and sufficient that $\mathfrak{L}_x = L_2(\mathbb{R}^1, d\sigma)$ or, in other words,

$$L_2(\mathbb{R}^1, d\sigma) = \text{c.l.s.}\{e^{i\lambda t}\}_{t\in(-\infty,0]}.$$

As was mentioned above (see the arguments after the proof of Theorem 2.3), this is possible if and only if

$$\int_{-\infty}^{\infty} \frac{\log \sigma'(\lambda)}{1 + \lambda^2} \, d\lambda = -\infty,$$

where $\sigma'(\lambda)$ is the derivative of the absolutely continuous part of the function $\sigma(\lambda)$. Thus we arrive at the following assertion.

Theorem 3.5 *In order that a unitary arc x_t, $t \in (-\infty, 0]$, admit a unique complete unitary continuation, it is necessary and sufficient that*

$$\int_{-\infty}^{\infty} \frac{\log \sigma'(\lambda)}{1 + \lambda^2} \, d\lambda = -\infty,$$

where $d\sigma(\lambda)$ is the measure in the representation (3.4) of the correlation function $F(t)$, $t \in \mathbb{R}^1$, of the arc x_t.

Theorem 3.6 *Assume that*

$$\int_{-\infty}^{\infty} \frac{\log \sigma'(\lambda)}{1 + \lambda^2} \, d\lambda > -\infty, \tag{3.5}$$

where $d\sigma(\lambda)$ is the same as in the previous theorem. A complete unitary continuation of the unitary arc x_t, $t \in (-\infty, 0]$, exists if and only if

$$\dim\left(\mathfrak{H} \ominus \mathfrak{L}_x\right) = \infty.$$

If that is the case, then the number of such continuations is infinite.

Proof. Let $\tilde{x}_t$, $t \in \mathbb{R}^1$, be a complete unitary continuation of the unitary arc x_t,
$t \in (-\infty, 0]$. Put

$$\mathfrak{L}^b = \text{c.l.s.}\left\{\tilde{x}_t\right\}_{t \in (-\infty, b]}.$$

Denote by $d\left(\tilde{x}_t, \mathfrak{L}^b\right)$ the distance from the element $\tilde{x}_t$ to the subspace $\mathfrak{L}^b$. As
Definition 3.4 shows, the equality

$$d\left(\tilde{x}_b, \mathfrak{L}^0\right) = d\left(\tilde{x}_{b-h}, \mathfrak{L}^{-h}\right)$$

is valid for any $h > 0$. So, if $\tilde{x}_a \in \mathfrak{L}^0$, $a > 0$, then $d\left(\tilde{x}_{a-h}, \mathfrak{L}^{-h}\right) = d\left(\tilde{x}_a, \mathfrak{L}^0\right) = 0$,
hence $d\left(\tilde{x}_{a-h}, \mathfrak{L}^0\right) = 0$ because $\mathfrak{L}^{-h} \subset \mathfrak{L}^0$. Thus, if $\tilde{x}_a \in \mathfrak{L}^0$, $a > 0$, then $\tilde{x}_t \in \mathfrak{L}^0$
for $t \leq a$, that is $\mathfrak{L}^a = \mathfrak{L}^0$. Since

$$d\left(\tilde{x}_{2a}, \mathfrak{L}^a\right) = d\left(\tilde{x}_a, \mathfrak{L}^0\right) = 0,$$

the element $\tilde{x}_{2a}$ belongs to $\mathfrak{L}^0$. Consequently all the line $\tilde{x}_t$ falls into $\mathfrak{L}^0$. Therefore,
if $\tilde{x}_a \notin \mathfrak{L}^0$, then the element $\tilde{x}_{na}$ does not belong to the space $\mathfrak{L}^{(n-1)a}$ ($n = 2, 3, \dots$). So the dimension of the subspace $\mathfrak{H} \ominus \mathfrak{L}^0 = \mathfrak{H} \ominus \mathfrak{L}_x$ is infinite.

Conversely, suppose $\dim\left(\mathfrak{H} \ominus \mathfrak{L}_x\right) = \infty$. The correspondence $x_t \mapsto e^{it\lambda}$,
$t \in (-\infty, 0]$, determines an isometric map from the subspace $\mathfrak{L}_x \subset \mathfrak{H}$ onto $L_\sigma^0 = \text{c.l.s.}\left\{e^{it\lambda}\right\}_{t \in (-\infty, 0]}$. Due to (3.5) we have $L_\sigma^0 \neq L_2\left(\mathbb{R}^1, d\sigma\right)$. But $e^{it\lambda}$, $t \in \mathbb{R}^1$, is a
unitary line in the space $L_2\left(\mathbb{R}^1, d\sigma\right)$. By the necessity condition established above,

$$\dim\left(L_2\left(\mathbb{R}^1, d\sigma\right) \ominus L_\sigma^0\right) = \infty.$$

Hence, there exists an operator U mapping isometrically $\mathfrak{H} \ominus \mathfrak{L}_x$ onto $L_2\left(\mathbb{R}^1, d\sigma\right) \ominus L_\sigma^0$ and $\mathfrak{L}^0 = \mathfrak{L}_x$ onto L_σ^0. It is evident that the line

$$\tilde{x}_t = U^{-1}e^{i\lambda t}, \quad t \in \mathbb{R}^1,$$

is a complete unitary continuation of the arc x_t.   □

3.3   Let x_t be a spiral arc. Without loss of generality we can consider it given on
the interval $[0, T]$. Then its metric function $B(s, t)$ is defined on $[-T, T]$.
It should be noted that the scalar product

$$\left(\sum_{i=1}^{n} \xi_i\, x_{s_i+r}\, , \, \sum_{i=1}^{n} \eta_k\, x_{t_k+r}\right)$$

does not depend on r if

$$\sum_{i=1}^{n} \xi_i = 0, \quad \sum_{i=1}^{n} \eta_k = 0 \,.$$

Indeed,

$$\left(\sum_{i=1}^{n} \xi_i x_{s_i+r}, \sum_{i=1}^{n} \eta_k\, x_{t_k+r} \right)$$

$$= \left(\sum_{i=1}^{n} \xi_i \left(x_{s_i+r} - x_r \right), \sum_{i=1}^{n} \eta_k \left(x_{t_k+r} - x_r \right) \right) = \sum_{i,k=1}^{n} B(s_i, t_k)\, \xi_i \overline{\eta}_k \,. \tag{3.6}$$

Corollary 3.1 *Let x_t, $t \in [0, T]$, be a spiral arc or a spiral in $\mathfrak{H}$. For any fixed $h \in [-\frac{T}{2}, \frac{T}{2}]$ the vector-function*

$$y_t = x_{t+h} - x_t, \quad t \in [0, \frac{T}{2}] \,,$$

is a unitary arc or a unitary line respectively. Conversely, if

$$x_{t+h} - x_t, \quad t \in [0, \frac{T}{2}] \,,$$

is a unitary arc or a unitary line for every $h \in [-\frac{T}{2}, \frac{T}{2}]$, then x_t, $t \in [0, T]$, is a spiral arc or a spiral.

Proof. Suppose x_t, $t \in [0, T]$, to be a spiral arc or a spiral. Then the scalar product

$$\begin{aligned}
(y_t, y_s) &= \left(x_{t+h} - x_t, x_{s+h} - x_s \right) = \left(x_{t-s+h} - x_{t-s}, x_h - x_0 \right) \\
&= B(t - s + h, h) - B(t - s, h) - B(t - s + h, 0) + B(t - s, 0)
\end{aligned}$$

depends only on the difference $t - s$. Hence, the arc or the line y_t is unitary.
    The converse is obvious.   □
    If a spiral x_t in $\mathfrak{H}$ is continuously differentiable, then

$$\lim_{h \to 0} \frac{1}{h^2} \left(x_{t+h} - x_t, x_{s+h} - x_s \right) = \left(x'_t, x'_s \right) = F(t - s) \,.$$

Thus, x'_t is a unitary line. Conversely, if x'_t is a unitary line, then x_t is a spiral. The latter follows from the equalities

$$\left(x_{r+s} - x_r, x_{r+t} - x_r \right) = \int_{r}^{r+s} \int_{r}^{r+t} \left(x'_\tau, x'_\sigma \right) d\tau\, d\sigma$$

$$= \int_{r}^{r+s} \int_{r}^{r+t} F(\tau - \sigma)\, d\tau\, d\sigma = \int_{0}^{s} \int_{0}^{t} F(\xi - \eta)\, d\xi\, d\eta = B(s, t) \,.$$

Note also that the Steklov average

$$x_t^h = \frac{1}{h} \int\limits_t^{t+h} x_s \, ds$$

of a spiral x_t is a continuously differentiable spiral, and the line

$$\frac{dx_t^h}{dt} = \frac{x_{t+h} - x_t}{h}$$

is unitary. Consequently, every spiral can be approximated by continuously differentiable ones.

3.4 Let a non-decreasing function $\tau(\lambda)$, $\lambda \in \mathbb{R}^1$, obey the normalization conditions

$$\tau(0) = 0, \quad \tau(\lambda) = \tau(\lambda - 0) . \tag{3.7}$$

Assume also that

$$\int\limits_{-\infty}^{\infty} \frac{d\tau(\lambda)}{1 + \lambda^2} < \infty . \tag{3.8}$$

Then the vector-function

$$\varphi_t = \frac{e^{i\lambda t} - 1}{\lambda}, \quad t \in \mathbb{R}^1 ,$$

is a spiral in the space $L_2(\mathbb{R}^1, d\tau)$. To show this find the scalar product $(\varphi_{r+s} - \varphi_r, \varphi_{r+t} - \varphi_r)$. We have

$$\left(\varphi_{r+s} - \varphi_r, \varphi_{r+t} - \varphi_r\right)$$

$$= \int\limits_{-\infty}^{\infty} \frac{\left(e^{i\lambda(r+s)} - 1\right) - \left(e^{i\lambda r} - 1\right)}{\lambda} \cdot \frac{\left(e^{-i\lambda(r+t)} - 1\right) - \left(e^{-i\lambda r} - 1\right)}{\lambda} \, d\tau(\lambda)$$

$$= \int\limits_{-\infty}^{\infty} \frac{\left(e^{i\lambda s} - 1\right)\left(e^{-i\lambda t} - 1\right)}{\lambda^2} \, d\tau(\lambda) .$$

As we can observe, this expression does not depend on r. So φ_t, $t \in \mathbb{R}^1$, is a spiral, and

$$B(s,t) = \int\limits_{-\infty}^{\infty} \frac{\left(e^{i\lambda s} - 1\right)\left(e^{-i\lambda t} - 1\right)}{\lambda^2} \, d\tau(\lambda), \quad s,t \in \mathbb{R}^1 ,$$

is its metric function.

Theorem 3.7 *Let x_t, $t \in [0, T]$, be a spiral arc in $\mathfrak{H}$. There exists at least one non-decreasing function $\tau(\lambda)$, $\lambda \in \mathbb{R}^1$, with properties (3.7) and (3.8), such that the metric function $B(s, t)$ of the arc x_t can be represented in the form*

$$B(s, t) = \int_{-\infty}^{\infty} \frac{\left(e^{i\lambda s} - 1\right)\left(e^{-i\lambda t} - 1\right)}{\lambda^2} \, d\tau(\lambda), \quad s, t \in [-T, T]. \tag{3.9}$$

Proof. Without loss of generality we may set $x_0 = 0$. Denote by $\mathfrak{L}_\omega^{(0)}$ the set of all generalized functions of the form

$$f(t) = \frac{d\omega_f(t)}{dt}$$

with compact supports within $(0, T)$, where $\omega_f(t)$ is a function of bounded variation (the derivative is taken in the Schwartz distribution sense). Besides, it is supposed that

$$\int_0^T f(t) \, dt = 0. \tag{3.10}$$

The bilinear form

$$(f, g)_x = \left(\int_0^T f(t)x_t \, dt, \int_0^T g(s)x_s \, ds \right)$$

determines, generally, a quasiscalar product in $\mathfrak{L}_\omega^{(0)}$. Here we mean the integral $\int_0^T f(t)x_t \, dt$ as

$$\int_0^T f(t)x_t \, dt = \int_0^T x_t \, d\omega_f(t).$$

Define the operator A_0 as

$$A_0 f = if',$$

$$\mathcal{D}(A_0) = \left\{ f \in \mathfrak{L}_\omega^{(0)} : f' \in \mathfrak{L}_\omega^{(0)} \right\}.$$

This operator is Hermitian. In fact, the set $\mathcal{D}(A_0)$ is quasidense in $\mathfrak{L}_\omega^{(0)}$, and if the arc x_t is continuously differentiable, then

$$
\begin{aligned}
\left(A_0 f, g\right)_x &= \left(i \int_0^T f'(t) x_t \, dt, \int_0^T g(s) x_s \, ds\right) \\
&= \left(-i \int_0^T f(t) \frac{dx_t}{dt} \, dt, \int_0^T g(s) x_s \, ds\right) \\
&= \lim_{h \to 0} \left(-i \int_0^T f(t) \frac{x_{t+h} - x_t}{h} \, dt, \int_0^T g(s) x_s \, ds\right) \\
&= \lim_{h \to 0} \left(\int_0^T f(t) x_t \, dt, \, i \int_0^T g(s) \frac{x_{s-h} - x_s}{h} \, ds\right) \\
&= \left(\int_0^T f(t) x_t \, dt, \, -i \int_0^T g(s) \frac{dx_s}{ds} \, ds\right) = \left(\int_0^T f(t) x_t \, dt, \, i \int_0^T g'(s) x_s \, ds\right) \\
&= (f, A_0 g)_x, \quad f, g \in \mathcal{D}(A_0) \, .
\end{aligned}
$$

In so doing, we have used the relation

$$
\left(\int_0^T f(t) x_{t+h} \, dt, \int_0^T g(s) x_s \, ds\right) = \left(\int_0^T f(t) x_t \, dt, \int_0^T g(s) x_{s-h} \, ds\right)
$$

which follows from the property (3.6) of the spiral arc x_t.

If the arc x_t is not differentiable, then the average

$$
x_t^h = \frac{1}{h} \int_t^{t+h} x_s \, ds
$$

is continuously differentiable. Therefore

$$
\left(A_0 f, g\right)_{x^h} = (f, A_0 g)_{x^h}, \quad f, g \in \mathcal{D}(A_0) \, .
$$

Taking into account that $x_t^h \to x_t$ as $h \to 0$, we obtain, by passing to the limit in the last identity, that the operator A_0 is Hermitian.

It is not difficult to verify that

$$
\Phi(f, \lambda) = \int_0^T \frac{e^{i\lambda t} - 1}{\lambda} f(t) \, dt = \int_0^T \frac{e^{i\lambda t} - 1}{\lambda} \, d\omega_f(t), \quad f \in \mathfrak{L}_\omega^{(0)}, \tag{3.11}
$$

is a directing functional for the operator A_0. Indeed, the solution of the equation

$$A_0 g - \lambda g = f, \quad f \in \mathcal{L}_\omega^{(0)},$$

can be written in the form

$$g(t) = -i\,e^{-i\lambda t} \int_0^t e^{i\lambda s} f(s)\,ds = -i\,e^{-i\lambda t} \int_0^t e^{i\lambda s}\,d\omega_f(s).$$

The condition $g(T) = 0$ yields

$$\int_0^T e^{i\lambda s} f(s)\,ds = 0,$$

which is sufficient for $g(t)$ to vanish in a certain neibourhood of the point $t = T$, because so does the function $f(t)$. Integrating termwise the equation

$$i g' - \lambda g = f,$$

we find, by virtue of (3.10), that

$$-\lambda \int_0^T g(s)\,ds = \int_0^T f(s)\,ds = 0,$$

whence

$$\int_0^T g(s)\,ds = 0, \quad \text{as } \lambda \neq 0 \qquad \text{and} \qquad g(t) = -i \int_0^t f(s)\,ds, \quad \text{as } \lambda = 0.$$

Since

$$\int_0^T g(t)\,dt = -i \int_0^T \int_0^t f(s)\,ds\,dt = -i \int_0^T (T-s) f(s)\,ds = 0,$$

the last condition in the case of $\lambda = 0$ implies

$$\int_0^T s f(s)\,ds = 0.$$

Thus, the function $\Phi(f, \lambda)$ given by the formula (3.11) is defined for any λ if we put

$$\frac{e^{i\lambda s} - 1}{\lambda} = is$$

for $\lambda = 0$.

By Theorem 2.8.1,

$$(f,g)_x = \int_{-\infty}^{\infty} \Phi(f,\lambda)\overline{\Phi(g,\lambda)}\, d\tau(\lambda), \quad f,g \in \mathcal{L}_\omega^{(0)}. \tag{3.12}$$

The substitution

$$f(\xi) = \delta_{t+r}(\xi) - \delta_r(\xi), \quad g(\xi) = \delta_{s+r}(\xi) - \delta_r(\xi)$$

in (3.12) gives

$$(f,g)_x = \left(x_{t+r} - x_r, x_{s+r} - x_r\right)$$

$$= \int_0^T \left(\frac{e^{i\lambda(t+r)}-1}{\lambda} - \frac{e^{i\lambda r}-1}{\lambda}\right) \cdot \left(\frac{e^{-i\lambda(s+r)}-1}{\lambda} - \frac{e^{-i\lambda r}-1}{\lambda}\right) d\tau(\lambda)$$

$$= \int_{-\infty}^{\infty} \frac{\left(e^{i\lambda t}-1\right)\left(e^{-i\lambda s}-1\right)}{\lambda^2}\, d\tau(\lambda).$$

So the representation (3.9) is true.

It remains only to check that

$$\int_{-\infty}^{\infty} \frac{d\tau(\lambda)}{1+\lambda^2} < \infty.$$

To this end take $s = t$ in (3.9). Since $x_0 = 0$, we have

$$\|x_t\|^2 = \int_{-\infty}^{\infty} \left|\frac{e^{i\lambda t}-1}{\lambda}\right|^2 d\tau(\lambda) = 2\int_{-\infty}^{\infty} \frac{1-\cos\lambda t}{\lambda^2}\, d\tau(\lambda). \tag{3.13}$$

Taking into account that $\left|\frac{\sin t}{t}\right| < \frac{1}{2}$ when $|t|$ is large enough, we get for $h > 0$ and sufficiently large N

$$\frac{1}{h}\int_0^h \|x_t\|^2\, dt = \frac{2}{h}\int_0^h \int_{-\infty}^{\infty} \frac{1-\cos\lambda t}{\lambda^2}\, d\tau(\lambda)\, dt$$

$$= \frac{2}{h}\int_{-\infty}^{\infty} \int_0^h \frac{1-\cos\lambda t}{\lambda^2}\, dt\, d\tau(\lambda) = \int_{-\infty}^{\infty} \frac{1-\frac{\sin h\lambda}{h\lambda}}{\lambda^2}\, d\tau(\lambda)$$

$$\geq \left(\int_{-\infty}^{-N} + \int_N^{\infty}\right) \frac{d\tau(\lambda)}{\lambda^2},$$

which implies (3.8). This completes the proof.   □

3.5   We now consider the case of $T < \infty$.

Denote by $\mathfrak{H}_x$ the completion of $\mathfrak{L}_\omega^{(0)}$ with respect to the quasiscalar product (3.12). The operator A_0 is real in the involution

$$J f(t) = \overline{f(T-t)} \,.$$

Since the functional (3.11) is universal, we can conclude in accordance with subsection 2.8.5 that the deficiency index of the operator A_0 is (0,0) or (1,1). It follows from the description (2.8.19) of the functions $\tau(\lambda)$ giving the representation (3.9) of the metric function $B(s,t)$ of the spiral arc x_t, that this representation is unique if and only if the operator A associated with A_0 as in subsection 2.8.2 is self-adjoint. Otherwise the measure $d\tau(\lambda)$ in (3.9) is not uniquely determined by the function $B(s,t)$.

If the element δ_0' belongs to the space $\mathfrak{H}_x$, then the operator A is entire with the ordinary gauge

$$u = \delta_0' \,, \qquad \text{because} \qquad \Phi(-i\delta_0', z) \equiv 1 \,, \quad z \in \mathbb{C}^1 \,.$$

In the opposite case the operator A is entire with respect to the generalized gauge

$$\tilde{u} = \delta_0' \,.$$

In both cases the distribution functions associated with the gauge u or $\tilde{u}$ of the operator A can be effectively described as has been done in sections 2.7 and 2.9. Recall that the canonical ones are step functions, and the jump points of different canonical distribution functions are alternating. Theorem 3.7 and the above arguments lead to the following assertion.

Theorem 3.8 *Every spiral arc x_t, $t \in [0,T]$, $T < \infty$, can always be continued to a complete spiral line $\tilde{x}_t$, $t \in \mathbb{R}^1$, whose values $\tilde{x}_t$ belong to the space*

$$\mathfrak{L}_x = \text{c.l.s. } \{x_t\}_{t \in [0,T]} \,.$$

The continuation is unique if and only if the representation (3.9) of the metric function $B(s,t)$ of the arc x_t, $t \in [0,T]$, is unique.

Proof.   Let the function $\tau(\lambda)$ be assigned to the given arc x_t, $t \in [0,T]$, by the formula (3.9), and the element

$$\varphi_t = \varphi_t(\lambda) = \frac{e^{it\lambda} - 1}{\lambda} \in L_2(\mathbb{R}^1, d\tau)$$

correspond to $x_t \in \mathfrak{L}_x$ in the isomorphism (1.2.6). Due to (3.9), this correspondence determines an isometric operator U from $\mathfrak{L}_x$ onto the subspace

$$L_\varphi = \text{c.l.s. } \left\{ \frac{e^{it\lambda} - 1}{\lambda} \right\}_{t \in [0,T]} \subset L_2(\mathbb{R}^1, d\tau) \,.$$

Choose the function $\tau(\lambda)$ so that

$$L_\varphi = L_2(\mathbb{R}^1, d\tau) \ .$$

If the deficiency index of A is $(0,0)$, then $\tau(\lambda)$ is unique; otherwise any canonical distribution function of the operator A may be taken as $\tau(\lambda)$. Denote by $\widetilde{\varphi}_t \in L_2(\mathbb{R}^1, d\tau)$ the natural continuation of φ_t from the interval $[0,T]$ to the whole real axis. Setting

$$\widetilde{x}_t = U^{-1} \widetilde{\varphi}_t \ ,$$

we obtain the complete spiral continuation $\widetilde{x}_t \in \mathfrak{L}_x$, $t \in \mathbb{R}^1$, of x_t, $t \in [0,T]$.

If $t \in \mathbb{R}^1$, then the function $\tau(\lambda)$ in (3.9) is uniquely determined by the metric function $B(s,t)$. This follows from the inversion formula for the Fourier transform and the fact (see, for instance, Gelfand and Shilov [1] or Schwartz [1]) that the derivative $-iF'(t)$ in the Schwartz distribution sense of the function

$$F(t) = \int\limits_{-\infty}^{\infty} \frac{e^{i\lambda t} - 1}{\lambda} \, d\tau(\lambda)$$

is the Fourier transform of the measure $d\tau(\lambda)$ whose growth order, in view of (3.8), is power.

In the nonuniqueness case in the representation (3.9), the deficiency index of the operator A is $(1,1)$. Choosing arbitrarily two different canonical distribution functions $\tau_1(\lambda)$ and $\tau_2(\lambda)$ of the operator A, we can construct two maps U_1 and U_2 which transform isometrically $\mathfrak{L}_x$ onto $L_2(\mathbb{R}^1, d\tau_1)$ and $L_2(\mathbb{R}^1, d\tau_2)$ respectively. Continuing in the natural way the spiral arc φ_t, $t \in [0,T]$, to spirals in the spaces $L_2(\mathbb{R}^1, d\tau_1)$ and $L_2(\mathbb{R}^1, d\tau_2)$ and then taking their inverse images in $\mathfrak{L}_x$, we get two complete spiral continuations of the arc x_t, $t \in [0,T]$. They are different because their metric functions are not identical.

Conversely, suppose that the spiral arc x_t, $t \in [0,T]$, has two different spiral continuations $\widetilde{x}_t^1$ and $\widetilde{x}_t^2$, $t \in \mathbb{R}^1$, in $\mathfrak{L}_x$. Then the metric functions $B_1(s,t)$ and $B_2(s,t)$ $(s,t \in \mathbb{R}^1)$ of these continuations do not coincide. Indeed, were this not the case, the equality $B_1(s,t) = B_2(s,t)$ would imply $\|\widetilde{x}_t^1\| = \|\widetilde{x}_t^2\|$, $t \in \mathbb{R}^1$, and therefore

$$\widetilde{x}_t^1 = V \, \widetilde{x}_t^2 \ ,$$

where

$$V : V \, x_t^1 = x_t^2$$

is a unitary operator on $\mathfrak{L}_x$. Since $\widetilde{x}_t^1 = \widetilde{x}_t^2$ as $t \in [0,T]$, $V = I$, whence $\widetilde{x}_t^1 = \widetilde{x}_t^2$, $t \in \mathbb{R}^1$. Hence $B(s,t)$ admits a nonunique representation of the form (3.9) on the interval $[0,T]$. □

Assume that a spiral arc x_t, $t \in [0,T]$, $T < \infty$, is not uniquely continued to a spiral. A complete spiral continuation taking its values in $\mathfrak{L}_x$ is called canonical. As we could see when proving Theorem 3.8, a complete spiral continuation $\widetilde{x}_t$,

$t \in \mathbb{R}^1$, of a spiral arc x_t, $t \in [0, T]$, is canonical if and only if the function $\tau(\lambda)$ appearing in the representation (3.9) of the metric function of the continuation $\widetilde{x}_t$, $t \in \mathbb{R}^1$, is a canonical distribution function of the operator A. It follows also from the reasoning in Theorem 3.7 that a certain distribution function $\tau(\lambda)$ of A corresponds to a complete spiral continuation $\widetilde{x}_t$, $t \in \mathbb{R}^1$, of a spiral arc x_t, $t \in [0, T]$. We shall say that this function $\tau(\lambda)$ is generated by the continuation $\widetilde{x}_t$. In the same manner as in the case of unitary arcs one can prove that if two different complete spiral continuations of a spiral arc generate one and the same distribution function of the operator A, then these continuations are unitary equivalent. If we suppose

$$\dim \mathfrak{H} \ominus \mathfrak{L}_x = \infty \,,$$

then for every distribution function $\tau(\lambda)$ of A there exists a certain class of unitary equivalent complete spiral continuations $\widetilde{x}_t$ of the spiral x_t which generate this $\tau(\lambda)$.

3.6   Let us clarify how a spiral x_t, $t \in \mathbb{R}^1$, behaves at ∞. By virtue of (3.13),

$$\|x_t\|^2 = 4 \int\limits_{-\infty}^{\infty} \frac{\sin^2 \frac{\lambda t}{2}}{\lambda^2} \, d\tau(\lambda) \,. \tag{3.14}$$

Put $m = \tau(+0) - \tau(0)$. Then

$$\|x_t\|^2 = mt^2 + 4 \int\limits_{-\infty}^{\infty} \frac{\sin^2 \frac{\lambda t}{2}}{\lambda^2} \, d\tau_0(\lambda) \,, \tag{3.15}$$

where the function $\tau_0(\lambda)$ has no jump at the point $\lambda = 0$. It is easily seen that the second summand in (3.15) is $o(t^2)$ near infinity. Therefore

$$\|x_t\|^2 = \big(m + o(1)\big)\, t^2 \quad (t \to \infty) \,.$$

The boundedness condition of a spiral is given in the next theorem.

Theorem 3.9 *In order that a spiral x_t, $t \in \mathbb{R}^1$, be bounded, it is necessary and sufficient that*

$$\int\limits_{-\infty}^{\infty} \frac{d\tau(\lambda)}{\lambda^2} < \infty \,, \tag{3.16}$$

where $d\tau(\lambda)$ is the measure in the representation (3.9) of the metric function $B(s, t)$ of the line x_t.

Proof. Obviously, the condition (3.16) implies, in view of (3.14), the boundedness of the spiral x_t.

Now let

$$\|x_t\| \leq R.$$

By the well-known Fatou theorem

$$R^2 \geq \frac{1}{h} \int\limits_0^h \|x_t\|^2 \, dt = 2 \int\limits_{-\infty}^{\infty} \frac{1 - \dfrac{\sin h\lambda}{h\lambda}}{\lambda^2} \, d\tau(\lambda)$$

$$\geq 2 \int\limits_{-N}^{N} \frac{1 - \dfrac{\sin h\lambda}{h\lambda}}{\lambda^2} \, d\tau(\lambda) \geq 2 \int\limits_{-N}^{N} \frac{d\tau(\lambda)}{\lambda^2},$$

which completes the proof since N is arbitrary. □

Theorem 3.10 *If a spiral x_t, $t \in \mathbb{R}^1$, is bounded, then x_t can be represented in the form*

$$x_t = c + y_t, \tag{3.17}$$

where y_t, $t \in \mathbb{R}^1$, is a unitary line, $c \in \mathfrak{L}_x$.

Proof. Let $d\tau(\lambda)$ be the measure in the representation (3.9) of the metric function $B(s,t)$ of the spiral x_t, which is uniquely determined by x_t. Then the correspondence

$$x_t \mapsto \frac{e^{i\lambda t} - 1}{\lambda}$$

gives an isometric isomorphism V from $\mathfrak{L}_x$ onto $L_2(\mathbb{R}^1, d\tau)$. Since the spiral x_t is bounded, we have

$$\int\limits_{-\infty}^{\infty} \frac{d\tau(\lambda)}{\lambda^2} < \infty.$$

Thus, $-\frac{1}{\lambda} \in L_2(\mathbb{R}^1, d\tau)$. Denote by c the inverse image of the function $-\frac{1}{\lambda}$ under the isomorphism V. The vector-function

$$u_t = \frac{e^{it\lambda}}{\lambda}, \quad t \in \mathbb{R}^1,$$

taking its values in the space $L_2(\mathbb{R}^1, d\tau)$ is a unitary line because the scalar product

$$(u_s, u_t)_{L_2(\mathbb{R}^1, d\tau)} = \int\limits_{-\infty}^{\infty} \frac{e^{is\lambda}}{\lambda} \cdot \frac{e^{-it\lambda}}{\lambda} \, d\tau(\lambda)$$

depends only on the difference $s - t$. Setting $y_t = V^{-1} u_t$, we conclude that the representation (3.17) is true. □

Theorem 3.11 *If a spiral arc x_t, $t \in [0, T]$, admits a nonunique continuation to a spiral, then the number of its canonical continuations is infinite. There is only one unbounded continuation among them. It is of the form*

$$\tilde{x}_t = at + y_t + c, \quad t \in \mathbb{R}^1,$$

where y_t is a unitary line whose values belong to $\mathfrak{L}_x$, $c \in \mathfrak{L}_x$, $a \in \mathfrak{L}_x$ is orthogonal to y_t for any $t \in \mathbb{R}^1$. All the rest are bounded, that is, of the form (3.17).

Proof. Let a spiral arc x_t, $t \in [0, T]$, be nonuniquely continued to a spiral. Then the deficiency index of the Hermitian operator A given on $\mathfrak{H}_x$ is $(1,1)$. So the number of its canonical distribution functions is infinite. Only one of these functions has a jump at the point $\lambda = 0$. Denote it by $\tau_0(\lambda)$. All the others are constant in a neighbourhood of zero. Hence,

$$\int\limits_{-\infty}^{\infty} \frac{d\tau(\lambda)}{\lambda^2} < \infty .$$

By Theorems 3.9 and 3.10 the complete spiral continuations $\tilde{x}_t$, $t \in \mathbb{R}^1$, of the arc x_t, $t \in [0, T]$, which correspond to all the functions $\tau(\lambda)$ except $\tau_0(\lambda)$ are bounded and represented in the form (3.17).

Consider the complete spiral continuation $\tilde{x}_t^0$ of the arc x_t, associated with the function $\tau_0(\lambda)$. The function

$$\theta_0(\lambda) = \begin{cases} 1 & \text{as} \ \lambda = 0 \\ 0 & \text{as} \ \lambda \neq 0 \end{cases}$$

is different from 0 in the space $L_2(\mathbb{R}^1, d\tau_0)$. Hence,

$$L_2(\mathbb{R}^1, d\tau_0) = \{\theta_0\} \oplus L_2(\mathbb{R}^1, d\tau_0') ,$$

where $\{\theta_0\}$ is the one-dimensional subspace spanned onto θ_0,

$$\tau_0'(\lambda) = \tau_0(\lambda) - m, \quad m = \tau_0(+0) - \tau_0(0) .$$

Then the spiral

$$y_t = \frac{e^{i\lambda t} - 1}{\lambda}$$

in the space $L_2(\mathbb{R}^1, d\tau_0)$ may be written in the form

$$\frac{e^{i\lambda t} - 1}{\lambda} = it\theta_0 + \left(\frac{e^{i\lambda t} - 1}{\lambda}\right)_{\tau_0'},$$

where

$$\left(\frac{e^{i\lambda t} - 1}{\lambda}\right)_{\tau_0'} = \begin{cases} 0 & \text{if } \lambda = 0 \\ \dfrac{e^{i\lambda t} - 1}{\lambda} & \text{if } \lambda \neq 0 \end{cases}$$

is a bounded spiral in $L_2(\mathbb{R}^1, d\tau_0') \subset L_2(\mathbb{R}^1, d\tau_0)$. The spiral $\tilde{x}_t^0$ is the inverse image of $\dfrac{e^{i\lambda t} - 1}{\lambda}$ under the isometric isomorphism V between $\mathfrak{L}_x$ and $L_2(\mathbb{R}^1, d\tau_0)$:

$$V^{-1}(i\theta_0)\, t + V^{-1}\left(\frac{e^{i\lambda t} - 1}{\lambda}\right)_{\tau_0'} = at + y_t + c,$$

which completes the proof.   □

3.7   It follows from the representation (3.9) that the metric function $B(s,t)$, $s, t \in [-T, T]$, of a spiral arc x_t, $t \in [0, T]$, can be represented as

$$B(s,t) = G(s - t) - G(s) - G(-t) + G(0),$$

where

$$G(t) = \int_{-\infty}^{\infty} \frac{e^{i\lambda t} - 1 - \dfrac{i\lambda t}{1 + \lambda^2}}{\lambda^2}\, d\tau(\lambda) + \alpha + i\beta t, \quad t \in [-2T, 2T].$$

Denote by $\mathfrak{G}_{2T}$ the class of all continuous Hermitian-symmetric functions $G(t)$ $(G(-t) = \overline{G(t)})$ given on the interval $[-2T, 2T]$, such that the kernel

$$B(s,t) = G(s - t) - G(s) - G(-t) + G(0)$$

is positive definite, i.e.

$$\sum_{j,k=1}^{n} B(t_j, t_k)\xi_j\overline{\xi}_k \geq 0, \quad t_j \in [-T, T], \quad j = 1, \ldots, n; \quad n \in \mathbb{N}.$$

Theorem 3.12 *In order that the function $G(t)$ belong to the class $\mathfrak{G}_{2T}$, it is necessary and sufficient that it admit an integral representation of the form*

$$G(t) = \alpha + i\beta t + \int_{-\infty}^{\infty} \frac{1}{\lambda^2}\left(e^{i\lambda t} - 1 - \frac{i\lambda t}{1 + \lambda^2}\right) d\tau(\lambda), \quad t \in [-2T, 2T], \quad (3.18)$$

where α, β are real constants, the function $\tau(\lambda)$ possesses the properties (3.7) and (3.8).

Proof. It is not difficult to verify that a continuous function $G(t)$ of the form (3.18) is Hermitian-symmetric and

$$B(s,t) = G(s-t) - G(s) - G(-t) + G(0) = \int\limits_{-\infty}^{\infty} \frac{e^{i\lambda s} - 1}{\lambda} \cdot \frac{e^{-i\lambda t} - 1}{\lambda} \, d\tau(\lambda) \,.$$

So,

$$\sum_{j,k=1}^{n} B(t_j, t_k) \xi_j \bar{\xi}_k = \int\limits_{-\infty}^{\infty} \left| \sum_{k=1}^{n} \frac{e^{i\lambda t_k} - 1}{\lambda} \xi_k \right|^2 d\tau(\lambda) \geq 0 \,.$$

Conversely, let a function $G(t)$ belong to the class $\mathfrak{G}_{2T}$. Denote by $\mathfrak{L}_\omega^{(0)}$ the set of all generalized functions of the form

$$f(t) = \frac{d\omega_f(t)}{dt}, \quad t \in [-T, T] \,,$$

with compact supports inside $(-T, T)$, where $\omega_f(t)$ is a function of bounded variation. We define a quasiscalar product in $\mathfrak{L}_\omega^{(0)}$ as

$$(f_1, f_2) = \int\limits_{-T}^{T} \int\limits_{-T}^{T} B(s,t) \, d\omega_{f_1}(t) \, \overline{d\omega_{f_2}(s)} \,. \tag{3.19}$$

Identifying the elements $f \in \mathfrak{L}_\omega^{(0)}$ such that $(f, f) = 0$ and completing with respect to (3.19), we obtain the Hilbert space $\mathfrak{H}_B$.

The vector-function

$$x_t = \delta_t(\xi), \quad 0 \leq t \leq T \,,$$

is a spiral arc in $\mathfrak{H}_B$. To show this, note first that if

$$\int\limits_{-T}^{T} d\omega_f(\xi) = \int\limits_{-T}^{T} \overline{d\omega_g(\eta)} = 0 \,, \quad \text{then} \quad (f, g) = \int\limits_{-T}^{T} \int\limits_{-T}^{T} G(\xi - \eta) \, d\omega_f(\xi) \, \overline{d\omega_g(\eta)} \,.$$

Since

$$\int\limits_{-T}^{T} \left(\delta_t(\xi) - \delta_r(\xi) \right) d\xi = 0 \,,$$

we have

$$(x_t - x_r, x_s - x_r) = \int\limits_{-T}^{T} \int\limits_{-T}^{T} G(\xi - \eta) \left(\delta_t(\xi) - \delta_r(\xi) \right) \left(\delta_s(\eta) - \delta_r(\eta) \right) d\xi \, d\eta$$

$$= G(t-s) - G(t-r) - G(r-s) + G(0) = B(t-r, s-r) \,.$$

Thus, $x_t = \delta_t(\xi)$, $t \in [0, T]$, is a spiral arc. Its metric function coincides with $B(s,t)$. By Theorem 3.7, $B(s,t)$ is represented in the form (3.9):

$$B(s,t) = \int_{-\infty}^{\infty} \frac{\left(e^{i\lambda s} - 1\right)\left(e^{-i\lambda t} - 1\right)}{\lambda^2}\, d\tau(\lambda),$$

where $\tau(\lambda)$:

$$\tau(0) = 0, \quad \tau(\lambda - 0) = \tau(\lambda)$$

is a non-decreasing function such that

$$\int_{-\infty}^{\infty} \frac{d\tau(\lambda)}{1 + \lambda^2} < \infty.$$

Using the function $\tau(\lambda)$, construct the function

$$G_0(t) = \int_{-\infty}^{\infty} \frac{1}{\lambda^2}\left(e^{i\lambda t} - 1 - \frac{i\lambda t}{1 + \lambda^2}\right) d\tau(\lambda).$$

One can check immediately that

$$B(s,t) = G_0(s - t) - G_0(s) - G_0(-t) + G_0(0).$$

Therefore, the function

$$H(t) = G(t) - G_0(t)$$

possesses the properties

$$H(t) = \overline{H(-t)}, \quad H(s - t) - H(s) - H(-t) + H(0) = 0, \quad s, t \in [-T, T].$$

Consequently, the continuous function

$$\mathcal{H}(t) = H(t) - H(0)$$

satisfies the functional equation

$$\mathcal{H}(s - t) = \mathcal{H}(s) + \mathcal{H}(-t), \quad s, t \in [-T, T],$$

whence

$$\mathcal{H}(t) = ct, \quad c = \text{const}.$$

The Hermitian symmetry of the function $\mathcal{H}(t)$ gives $c = i\beta$, $\Im\beta = 0$. Put $\alpha = H(0)$. Then

$$H(t) = \mathcal{H}(t) + H(0) = i\beta t + \alpha,$$

and hence,

$$G(t) = G_0(t) + \alpha + i\beta t, \quad \Im\alpha = \Im\beta = 0$$

which completes the proof. $\square$

Remark 3.1 *If*

$$\int\limits_{-\infty}^{\infty} \frac{d\tau(\lambda)}{\lambda^2} < \infty \, ,$$

then the formula (3.18) *may be reduced, by choosing* α *and* β, *to the form*

$$G(t) = \int\limits_{-\infty}^{\infty} \frac{e^{it\lambda}}{\lambda^2} \, d\tau(\lambda) \, .$$

The next theorem follows from Theorem 3.12.

Theorem 3.13 *The function* $B(s,t)$, $s, t \in [-T, T]$, *is the metric function of a certain spiral arc if and only if it admits the representation*

$$B(s,t) = G(s - t) + G(s) - G(-t) + G(0) \, ,$$

where the function $G(t)$ *belongs to the class* $\mathfrak{G}_{2T}$.

Theorem 3.12 enables us to conclude that every function $G(t) \in \mathfrak{G}_{2T}, : T < \infty$, can be continued in the natural way (namely by the expression (3.18)) into the class $\mathfrak{G}_{\infty}$.

In the case of $T = \infty$, the measure $d\tau(\lambda)$ in the representation (3.18) is uniquely determined by the function $G(t)$. This is not the case if $T < \infty$. In the latter situation there exists a one-to-one correspondence between the functions $\tau(\lambda)$ appearing in the representation (3.18) and the continuations to the whole real axis from the class $\mathfrak{G}_{\infty}$ of the function $G(t) \in \mathfrak{G}_{2T}$. Hence, the description problem of all continuations of the function $G(t) \in \mathfrak{G}_{2T}$ into the class $\mathfrak{G}_{\infty}$ is equivalent to that of all the measures $d\tau(\lambda)$ in the representation (3.18). As has been established in subsection 2.9.5, these measures are given by the formula (2.9.22) connecting them in a one-to-one manner with the spectral functions of the entire operator A defined in the proof of Theorem 3.7.

Appendix 1

On Entire Operators whose
Defect Numbers are Arbitrary

The principal aspects of the theory of Hermitian operators whose defect numbers (finite or infinite) are arbitrary, entire with respect to an ordinary as well as a generalized gauge, are presented in this Appendix.

1.1 In what follows we shall need some results on integration of operator-functions with respect to an operator function of bounded variation.

Let $\mathfrak{H}_1$ and $\mathfrak{H}_2$ be separable Hilbert spaces. A function $\sigma(t)$ given on the interval $[a, b]$ with values in the set $L[\mathfrak{H}_1, \mathfrak{H}_2]$ of all bounded operators acting from $\mathfrak{H}_1$ into $\mathfrak{H}_2$ is called a function of weakly bounded variation if for any $f \in \mathfrak{H}_1, g \in \mathfrak{H}_2$ the scalar function $(\sigma(t)f, g)$ has a bounded variation on $[a, b]$. We denote the set of all such functions $\sigma(t)$ by $V_a^b(\mathfrak{H}_1, \mathfrak{H}_2)$. As has been shown in Danford and Schwartz [1],

$$\|\sigma\|_V = \sup_{\|f\|=1, \|g\|=1} \operatorname*{Var}_{[a,b]} (\sigma(t)f, g) < \infty.$$

for any $\sigma(t) \in V_a^b(\mathfrak{H}_1, \mathfrak{H}_2)$.

Theorem 1.1 *Let $\sigma(t) \in V_a^b(\mathfrak{H}_1, \mathfrak{H}_2)$, $\varphi(t)$ and $\psi(t)$ be analytic functions on $[a, b]$ taking their values in $L[\mathfrak{H}_2]$ and $L[\mathfrak{H}_1]$ respectively $(L[\mathfrak{H}] = L[\mathfrak{H}, \mathfrak{H}])$. Then the integral*

$$\int_a^b \varphi(t)\, d\sigma(t)\, \psi(t) \tag{1.1}$$

exists, which is understood as a limit of the Riemann-Stieltjes sums.
If

$$\varphi(t) = \sum_{k=0}^{\infty} \varphi_k (t - t_0)^k, \quad \psi(t) = \sum_{k=0}^{\infty} \psi_k (t - t_0)^k$$

are the Taylor expansions of the functions $\varphi(t)$ and $\psi(t)$, convergent on the interval $[t_0 - r, t_0 + r] \supset [a, b]$, then

$$\left\| \int_a^b \varphi(t)\, d\sigma(t)\, \psi(t) \right\| \leq \|\sigma\|_V \cdot \|\varphi\|_r \cdot \|\psi\|_r,$$

where

$$\|\varphi\|_r = \sum_{k=0}^{\infty} \|\varphi_k\| \, r^k \; (< \infty).$$

This theorem is proved in the same way as the corresponding one in Langer's paper [1], where the case of a Hermitian non-decreasing $\sigma(t)$ is considered. Moreover, the example is given there when the integral (1.1) does not exist if the functions $\varphi(t)$ and $\psi(t)$ are only continuous.

Denote by $\mathfrak{S}^k$, $k = 0, 1, 2, \ldots$, the class of all scalar complex measures $\tau(\Delta)$ given on Borel sets $\Delta \subset \mathbb{R}^1$ such that

$$\int_{-\infty}^{\infty} \frac{|d\tau(\lambda)|}{1 + |\lambda|^k} < \infty. \tag{1.2}$$

In particular, $\mathfrak{S}^0$ is the set of all measures having a bounded variation on the whole real axis.

We say that an operator measure $\sigma(\Delta)$ (i.e. a countably additive function of Δ) with values in $L[\mathfrak{H}_1, \mathfrak{H}_2]$ belongs to the class $\mathfrak{S}^k(\mathfrak{H}_1, \mathfrak{H}_2)$ if for any $f \in \mathfrak{H}_1, g \in \mathfrak{H}_2$ the scalar measure $(\sigma(\Delta)f, g) \in \mathfrak{S}^k$, i.e. satisfies (1.2).

Suppose $\sigma(\Delta) \in \mathfrak{S}^1(\mathfrak{H}_1, \mathfrak{H}_2)$. Then the Cauchy integral

$$C_\sigma(\lambda) = \int_{-\infty}^{\infty} \frac{d\sigma(t)}{t - \lambda} \tag{1.3}$$

exists. The operator function $C_\sigma(\lambda)$ takes its values in $L[\mathfrak{H}_1, \mathfrak{H}_2]$. It is analytic inside the upper and lower half-planes and also on the intervals of constancy of the function $\sigma(\lambda)$. The Stieltjes inversion formula (see Theorem 1.2.4) is carried in the ordinary way over to the operator case. Its operator analogue can be written as

$$\frac{1}{2\pi i} \oint_{\Gamma_{ab}}' C_\sigma(\lambda) \, d\lambda = -\int_a^b d\bar{\sigma}(t), \tag{1.4}$$

where the integration contour Γ_{ab} on the left-hand side of (1.4) intersects the real axis at the points a and b at a nonzero angle, the integral is taken in the principal value sense, $\bar{\sigma}(t) = \frac{1}{2}\big(\sigma(t - 0) + \sigma(t + 0)\big)$.

Now let $[a, b]$ be a closed interval of the real axis. The formula (1.4) is easily extended to the functions of the form

$$\Phi(\lambda) = \Psi(\lambda) + \int_{a'}^{b'} \frac{d\tau(t)}{t - \lambda}. \tag{1.5}$$

Here the operator-function $\Psi(\lambda)$ with values in $L[\mathfrak{H}_1, \mathfrak{H}_2]$ is analytic on $[a, b]$, $[a', b'] \supset [a, b]$, $\tau(\Delta) : \Delta \to L[\mathfrak{H}_1, \mathfrak{H}_2]$ is an operator measure on $[a', b']$. If $\Phi(\lambda)$

admits the representation (1.5), then for any $[c, d] \subset [a, b]$,

$$\frac{1}{2\pi i} \oint_{\Gamma_{cd}}' \Phi(\lambda) \, d\lambda = -\int_c^d d\overline{\tau}(t) \, . \tag{1.6}$$

The next assertion follows from the formula (1.6).

Theorem 1.2 *In order that the function* $\Phi(\lambda)$ *of the form* (1.5) *be analytic on the interval* $[a_1, b_1] \subset [a, b]$, *it is necessary and sufficient that for an arbitrary interval* $[c, d] \subset [a_1, b_1]$

$$\oint_{\Gamma_{cd}}' \Phi(\lambda) \, d\lambda = 0 \, .$$

If for any f *from a certain set* $\mathfrak{L}$ *dense in* $\mathfrak{H}$ *the vector-function* $\Phi(\lambda)f$ *admits an analytic continuation to the interval* $[a, b]$, *then the operator-function* $\Phi(\lambda)$ *is also analytically continuable to* $[a, b]$.

In the case where either $\mathfrak{H}_1$ or $\mathfrak{H}_2$ is one-dimensional, we obtain the following statement concerning vector-functions which are represented by the Cauchy integral.

Corollary 1.1 *Suppose that for an arbitrary vector* f *from a dense set* $\mathfrak{L} \in \mathfrak{H}$ *the scalar function* $(\Phi(\lambda), f)$, *where the vector-function* $\Phi(\lambda)$ *with values in* $\mathfrak{H}$ *is of the form* (1.5), *can be analytically continued to the interval* $[a, b]$. *Then* $\Phi(\lambda)$ *is analytically continuable to* $[a, b]$

The following lemma gives a condition sufficient for a function to admit the representation of the form (1.5).

Lemma 1.1 *Let* $C_\sigma(\lambda)$ *be an operator-function of the form* (1.3), $\varphi(\lambda)$ *and* $\psi(\lambda)$ *functions analytic on* $[a, b]$ *with values in* $L[\mathfrak{H}'_1, \mathfrak{H}_1]$ *and* $L[\mathfrak{H}_2, \mathfrak{H}'_2]$ *respectively. Then the function* $\psi(\lambda)C_\sigma(\lambda)\varphi(\lambda)$ *can be represented in the form* (1.5).

Proof. Choose an interval (a', b') so that $[a, b] \subset (a', b')$ and the functions $\varphi(\lambda)$ and $\psi(\lambda)$ are analytic on (a', b'). Then

$$\psi(\lambda)C_\sigma(\lambda)\varphi(\lambda) = \psi(\lambda)\left(\int_{-\infty}^{a'} + \int_{a'}^{b'} + \int_{b'}^{\infty}\right) \frac{d\sigma(t)}{t - \lambda} \varphi(\lambda)$$

$$= \psi(\lambda)\left(\int_{-\infty}^{a'} + \int_{b'}^{\infty}\right) \frac{d\sigma(t)}{t - \lambda} \varphi(\lambda) - \int_{a'}^{b'} \frac{\psi(t) - \psi(\lambda)}{t - \lambda} d\sigma(t) \varphi(\lambda)$$

$$- \int_{a'}^{b'} \psi(t) \, d\sigma(t) \frac{\varphi(t) - \varphi(\lambda)}{t - \lambda} + \int_{a'}^{b'} \frac{\psi(t) \, d\sigma(t) \varphi(t)}{t - \lambda} \, ,$$

whence

$$\psi(\lambda)C_\sigma(\lambda)\varphi(\lambda) = \Psi(\lambda) + \int\limits_{a'}^{b'} \frac{d\tau(\lambda)}{t-\lambda}, \qquad (1.7)$$

where the function $\Psi(\lambda)$ is analytic on $[a, b]$ and

$$\tau(\Delta) = \int\limits_{\Delta} \psi(t)\, d\sigma(t)\, \varphi(t). \qquad \square$$

Applying the inversion formula (1.6) to (1.7), we get

$$\frac{1}{2\pi i} \oint\limits_{\Gamma_{cd}}' \psi(\lambda)C_\sigma(\lambda)\varphi(\lambda)\, d\lambda = -\int\limits_{c}^{d} \psi(t)\, d\overline{\sigma}(t)\, \varphi(t), \quad (c, d) \subset [a, b].$$

This formula is a generalization of (1.2.9).

1.2 Let A, $\overline{\mathcal{D}(A)} = \mathfrak{H}$ be a simple closed Hermitian operator whose deficiency index is (n, n), $n \leq \infty$, and M a subspace of $\mathfrak{H}$ (so-called gauge).

Definition 1.1 *A point $z \in \mathbb{C}^1$ is called M-regular for the operator A if*

$$\mathfrak{M}_z = (A - zI)\mathcal{D}(A) = \overline{\mathfrak{M}_z} \quad \text{and} \quad \mathfrak{H} = \mathfrak{M}_z \dotplus M.$$

It follows from the simplicity and closedness of the operator A and the closedness of $\mathfrak{M}_z$ that every M-regular point of A is its point of regular type. As has been shown by Gohberg and Markus [1], the set $\rho(A, M)$ of all M-regular points of the operator A is open.

Suppose $z \in \rho(A, M)$ and denote by $P(z)$ the projector from $\mathfrak{H}$ onto M parallel to $\mathfrak{M}_z$. So, for any $f \in \mathfrak{H}$ the vector $f - P(z)f$ belongs to $\mathfrak{M}_z$. Hence, there exists a unique element $g \in \mathcal{D}(A)$ such that

$$f - P(z)f = Ag - zg.$$

We introduce the operator $T(z)$ by setting

$$T(z)f = g,$$

that is,

$$T(z) = (A - zI)^{-1}(I - P(z)). \qquad (1.8)$$

The relation (1.8) shows that $T(z)$ is a bounded operator on $\mathfrak{H}$. Thus, we deal with two functions $P(z)$ and $T(z)$ defined on the set $\rho(A, M)$ whose values belong to $L[\mathfrak{H}]$. They are called functions of the first and the second kind, respectively, associated with the gauge M. These functions possess the properties:

$1°$ $P(z)Af = zP(z)f, \quad f \in \mathcal{D}(A)$.

$2°$ $P(z)P(\zeta) = P(\zeta), \quad z, \zeta \in \rho(A, M)$.

$3°$ a) $T(z)M = 0$;

   b) $T(z)\mathfrak{H} = \mathcal{D}(A)$;

   c) $(A - zI)T(z) = I - P(z)$;

   d) $T(z)(A - zI)g = g, \quad g \in \mathcal{D}(A)$.

Indeed, since $f = Ag - zg \in \mathfrak{M}_z$, we have $P(z)f = 0$. Hence,

$$f - P(z)f = f.$$

The property $1°$ implies

$$T(z)(Ag - zg) = T(z)f = (A - zI)^{-1}(I - P(z))f = (A - zI)^{-1}f = g.$$

$4°$ $P(z) - P(\zeta) = (z - \zeta)P(\zeta)T(z), \quad z, \zeta \in \rho(A, M)$.

In fact, according to the properties $1°$, $2°$, and $3°$ c),

$$(z - \zeta)P(\zeta)T(z) = P(\zeta)(zI - A)T(z) = -P(\zeta)(I - P(z))$$
$$= P(\zeta)P(z) - P(\zeta) = P(z) - P(\zeta).$$

$5°$ The function $T(z)$ satisfies the resolvent equation

$$T(z) - T(\zeta) = (z - \zeta)T(z)T(\zeta), \quad z, \zeta \in \rho(A, M). \tag{1.9}$$

Indeed,

$$P(z) - P(\zeta) = (z - \zeta)P(\zeta)T(z) + (z - \zeta)T(z) - (z - \zeta)T(z)$$
$$= -(z - \zeta)(I - P(\zeta))T(z) + (z - \zeta)T(z),$$

whence

$$(z - \zeta)T(z) - P(z) + P(\zeta) = (z - \zeta)(I - P(\zeta))T(z),$$

i.e.

$$-I + P(z) + (A - \zeta I)(A - zI)^{-1}(I - P(z)) - P(z) + P(\zeta) = (z - \zeta)(I - P(\zeta))T(z).$$

Hence,

$$-I + (A - \zeta I)(A - zI)^{-1}(I - P(z)) + P(\zeta) = (z - \zeta)(I - P(\zeta))T(z).$$

Applying the operator $(A - \zeta I)^{-1}$ to both parts of the last equality, we obtain

$$T(z) - T(\zeta) = (z - \zeta)T(\zeta)T(z).$$

6° The function $T(z)$ is analytic on $\rho(A, M)$ and all its values commute with each other.

7° The function $\mathcal{P}(z)$ is analytic on $\rho(A, M)$.

In the more general situation of a closed operator in a Banach space these properties were established by Gohberg and Markus [1].

The functions $\mathcal{P}(z)$ and $T(z)$ are a priori defined on $\rho(A, M)$. The next theorem shows that $\rho(A, M)$ is the natural analyticity set of $T(z)$, that is, the function $T(z)$ does not admit an analytic continuation to any wider set.

Theorem 1.3 *Suppose that the function $T(z)$ can be analytically continued to the domain $\tilde{\rho} \supset \rho(A, M)$. Then the set $\tilde{\rho}$ is univalent and consists of M-regular points of the operator A.*

Proof. The function $T(z)$ being a solution of the Hilbert resolvent equation (1.9) is a one-valued analytic function. Hence, the domain $\tilde{\rho}$ is univalent. It follows from the properties 4° and 3° c) that $\mathcal{P}(z)$ admits an analytic continuation to $\tilde{\rho}$ without loss of the property of being a one-valued analytic function. The permanence principle allows one to conclude that this continuation possesses all the properties $1° - 7°$.

By virtue of the property 3° d),

$$\|g\| \leq \|T(z)\| \cdot \|(A - zI)g\|, \quad z \in \tilde{\rho}.$$

So,
$$\|(A - zI)g\| \geq \frac{1}{\|T(z)\|} \|g\|, \quad z \in \tilde{\rho}.$$

Thus, $z \in \tilde{\rho}$ is a point of regular type for the operator A.

It is evident that the property that the operator $\mathcal{P}(z)$ is the projector from $\mathfrak{H}$ onto M is kept for its analytic continuation. It is also obvious that, due to the property 1°,

$$\mathcal{P}(z)(Af - zf) = 0, \quad f \in \mathcal{D}(A), \quad z \in \tilde{\rho},$$

which is equivalent to

$$\mathcal{P}(z)\mathfrak{M}_z = 0, \quad z \in \tilde{\rho}.$$

It remains to prove that the converse is valid, too, i.e. the equality $\mathcal{P}(z)u = 0$ implies $u \in \mathfrak{M}_z$. If we show this, then $I - \mathcal{P}(z)$ will be the projector from $\mathfrak{H}$ onto $\mathfrak{M}_z$, therefore

$$\mathfrak{H} = M \dotplus \mathfrak{M}_z, \quad z \in \tilde{\rho}.$$

But as is known, the range of the function $T(z)$ satisfying the Hilbert resolvent equation does not depend on z. So,

$$T(z)\mathfrak{H} = \mathcal{D}(A) \quad \text{for all } z \in \tilde{\rho}.$$

Put
$$x = T(z)u.$$

It follows from 3° c) that

$$(A - zI)x = (A - zI)T(z)u = (I - \mathcal{P}(z))u = u.$$

Consequently, $u \in \mathfrak{M}_z$. □

From what has been said we conclude that

$$\tilde{\rho} = \rho(A, M).$$

Lemma 1.2   *If R_λ is a generalized resolvent of the operator A, then for any non-real numbers λ and μ, the operator*

$$U_{\lambda\mu} = I + (\lambda - \mu)R_\lambda$$

transforms $\mathfrak{M}_\lambda$ bijectively onto $\mathfrak{M}_\mu$.

Proof. Set

$$f = (A - \lambda I)g, \quad g \in \mathcal{D}(A).$$

We have

$$U_{\lambda\mu}f = (I + (\lambda - \mu)R_\lambda)(Ag - \lambda g) = (Ag - \lambda g) + (\lambda - \mu)g = Ag - \mu g,$$

i.e. $U_{\lambda\mu}f \in \mathfrak{M}_\mu$. The one-to-one-ness of the mapping $U_{\lambda\mu} : \mathfrak{M}_\lambda \to \mathfrak{M}_\mu$ follows from the fact that every non-real point is a point of regular type for the operator A. □

Theorem 1.4 *If for a vector $f_0 \in \mathfrak{H}$ the vector-function $\mathcal{P}(z)f_0$ admits an analytic continuation to a closed interval Δ_0 of the real axis, then:*

(i) *the vector-function $T(z)f_0$ can be analytically continued to Δ_0;*

(ii) *for any spectral function E_λ of the operator A*

$$E(\Delta_0)f_0 = \int_{\Delta_0} dE_t \mathcal{P}(t)f_0. \tag{1.10}$$

Proof. (i) Let R_z be a generalized resolvent of A. It follows from the equality

$$R_z = \int_{-\infty}^{\infty} \frac{dE_\lambda}{\lambda - z} \tag{1.11}$$

and Lemma 1.1 applied to the case of $\mathfrak{H}_1 = \mathfrak{H}_2 = \mathfrak{H}'_1 = \mathfrak{H}$, $\mathfrak{H}'_1 = \{f_0\}$, $\psi(\lambda) \equiv I$, $\varphi(\lambda) = I - \mathcal{P}(\lambda)$ that the vector-function $T(\lambda)f_0$ is represented in the form (1.5). According to Corollary 1.1, it suffices to prove that the scalar functions $(T(z)f_0, g)$ are analytically continuable to Δ_0 for all g from a set $\mathcal{L}$ dense in $\mathfrak{H}$. We choose the linear span of all subspaces $\mathfrak{N}_z$, $\Im z \neq 0$, as $\mathcal{L}$. Since the operator A is simple, this linear span is dense in $\mathfrak{H}$.

So, assume $g \in \mathfrak{N}_{\bar{\mu}}$ for some $\mu : \Im\mu \neq 0$. Then

$$T(z) = \frac{U_{z\mu} - I}{z - \mu}(I - \mathcal{P}(z)) = -\frac{I - \mathcal{P}(z)}{z - \mu} + \frac{U_{z\mu}(I - \mathcal{P}(z))}{z - \mu}.$$

By Lemma 1.2, $U_{z\mu}(I - \mathcal{P}(z))\mathfrak{H} \subset \mathfrak{M}_\mu$. Therefore

$$(T(z)f_0, g) = -\frac{((I - \mathcal{P}(z))f_0, g)}{z - \mu}.$$

This equality shows that the function $(T(z)f_0, g)$ admits an analytic continuation to Δ_0.

(ii)   Let R_z be the generalized resolvent associated with the spectral function E_Δ of the operator A by the formula (1.11). Applying Theorem 1.2 to the vector-function $T(\lambda)f_0 = R_\lambda(I - \mathcal{P}(\lambda))f_0$, we obtain

$$\int_{\Delta_0} dE_\lambda(I - \mathcal{P}(\lambda))f_0 = 0,$$

which is equivalent to the equality (1.10). $\square$

Corollary 1.2 *If the function $\mathcal{P}(z)$ admits an analytic continuation to a closed interval Δ_0 of the real axis, so does the function $T(z)$, hence the interval Δ_0 consists of M-regular points of the operator A. In particular, every point of the interval Δ_0 is a point of regular type for A.*

Define the operator measure $\sigma(\Delta)$ in M by the relation

$$(\sigma(\Delta)f, g) = (E_\Delta f, g), \quad f, g \in M, \tag{1.12}$$

for any closed interval $\Delta \subset \mathbb{R}^1$, where E_λ is a spectral function of the operator A. This measure is positive (its values are non-negative operators on M) and finite (for any $f \in M$ the function $(\sigma(t)f, f)$ is bounded).

Theorem 1.5 *If for some $f, g \in \mathfrak{H}$ the vector-functions $\mathcal{P}(z)f$ and $\mathcal{P}(z)g$ can be analytically continued to the interval $\Delta_0 = [a, b] \subset \mathbb{R}^1$, then*

$$(E_{\Delta_0}f, g) = \int_{\Delta_0} (d\sigma(t)\mathcal{P}(t)f, \mathcal{P}(t)g). \tag{1.13}$$

Proof. In view of (1.10),

$$(E_{\Delta_0}f, g) = \int_{\Delta_0} (dE_t\mathcal{P}(t)f, g) = \int_{\Delta_0} (\mathcal{P}(t)f, dE_t g). \tag{1.14}$$

On the other hand, writing (1.10) for the interval $[a, t]$, we get

$$E_{[a,t]}g = \int_a^t dE_s\mathcal{P}(s)g.$$

The substitution of this expression in (1.14) yields

$$(E_{\Delta_0} f, g) = \int_{\Delta_0} (P(t)f, dE_t P(t)g) = \int_{\Delta_0} (d\sigma(t)P(t)f, P(t)g),$$

which is what had to be proved.   □

1.3 Suppose A to be a simple closed operator on $\mathfrak{H}$ whose defect numbers are equal.

Definition 1.2 *The operator A is called entire (M-entire) if the set $\rho(A, M)$ of all its M-regular points coincides with the whole complex plane.*

If the operator A is M-entire, then the operator-functions $P(z)$ and $T(z)$ with values in $L[\mathfrak{H}]$ are entire.

Let the entire function

$$f_M(z) = P(z)f$$

taking its values in M correspond to a vector $f \in \mathfrak{H}$. By Theorem 1.2.1, the map

$$f \mapsto f_M(z) \tag{1.15}$$

from $\mathfrak{H}$ into a certain subspace of entire functions with values in M is one-to-one. The map (1.15) transforms the operator A into the multiplication by z, that is

$$g_M(z) = z f_M(z) \quad \text{if} \quad g = Af.$$

It follows from Theorem 1.5 that for any $f, g \in \mathfrak{H}$

$$(f, g) = \int_{-\infty}^{\infty} (d\sigma(t)f_M(t), g_M(t)), \tag{1.16}$$

where the operator measure $\sigma(\Delta)$ in M is defined by the relation (1.12). Repeating word for word the proof of sufficiency in Theorem 2.1.1 for the partial case of $\mathfrak{R} = \mathfrak{H}$ and taking into account that $P(z)f = f$ as $f \in M$, we arrive at the following conclusion: if for a finite positive operator measure in M the equality (1.16) holds for any $f, g \in \mathfrak{H}$, then there exists a spectral function of the operator A such that $\sigma(\Delta)$ is represented in the form (1.12). Thus, in the case where the operator A is entire the set of all measures satisfying the condition (1.16) is completely classified by the formula (1.12) in which E_λ runs through the set of all spectral functions of the operator A. It turns out that all such measures $\sigma(\Delta)$ can be described in a way analogous to that used in section 2.7 for an entire operator whose deficiency index is (1,1).

Put

$$\mathcal{J} = i\mathcal{J}_0 = i \begin{pmatrix} 0 & -I_M \\ I_M & 0 \end{pmatrix}, \quad G(z) = \begin{pmatrix} P(z) \\ -Q(z) \end{pmatrix}, \tag{1.17}$$

where $Q(z) = P_M T(z), \quad z \in \mathbb{C}^1,$ P_M is the orthoprojector from $\mathfrak{H}$ onto M.

Theorem 1.6 *There exists a matrix-function*

$$W(z) = \begin{pmatrix} W_{11}(z) & W_{12}(z) \\ W_{21}(z) & W_{22}(z) \end{pmatrix}$$

whose entries are entire operator-functions with values in $L[M]$ such that for any $z, \zeta \in \mathbb{C}^1$

$$W(z)\mathcal{J}W^*(\zeta) = \mathcal{J} + \frac{1}{i}(z - \bar{\zeta})G(z)G^*(\zeta)\big|_M$$

$$= -i \begin{pmatrix} (z - \bar{\zeta})\mathcal{P}(z)\mathcal{P}^*(\zeta)\big|_M & I_M - (z - \bar{\zeta})\mathcal{P}(z)\mathcal{Q}^*(\zeta)\big|_M \\ -I_M - (z - \bar{\zeta})\mathcal{Q}(z)\mathcal{P}^*(\zeta)\big|_M & (z - \bar{\zeta})\mathcal{Q}(z)\mathcal{Q}^*(\zeta)\big|_M \end{pmatrix}. \tag{1.18}$$

The matrix $W(z)$ is invertible in the space $L[M \oplus M]$. (Here W^* denotes the matrix which is obtained from W by transposing and next changing all its components by their adjoints.)

It should be noted that the function $W(z)$ is determined by (1.18) up to the transform

$$W(z) \mapsto W(z)\mathcal{E}$$

where $\mathcal{E}$ is a constant $\mathcal{J}$-unitary matrix from the space $L[M \oplus M]$. Recall that the matrix $\mathcal{E}$ is called $\mathcal{J}$-unitary if it is invertible and $\mathcal{E}^*\mathcal{J}\mathcal{E} = \mathcal{J}$.

The formula (1.18) shows that the matrix $W(z)$ is $\mathcal{J}$-expanding as $\Im z > 0$ and $\mathcal{J}$-contracting as $\Im z < 0$. If $a \in \mathbb{R}^1$, the matrices $W^*(a)$ and $W(a)$ are $\mathcal{J}$-unitary.

For a real number a the matrix-function $W(z)$ may be chosen so that $W(a) = I_{M \oplus M}$. Setting $\zeta = a$ in (1.18), we get

$$W(z) = I_{M \oplus M} + (z - a)G(z)G^*(a)\mathcal{J}_0 \tag{1.19}$$

($\mathcal{J}_0$ and G are defined by (1.17)). If the operator A is real with respect to an involution J and the subspace M is invariant with respect to J, then the matrix-function $W(z)$ in (1.19) can be selected so that it is real, i.e.

$$W(\bar{z}) = JW(z), \quad \text{and symplectic, i.e.} \quad W'(z)\mathcal{J}W(z) = \mathcal{J}$$

(' denotes the transposition sign).

Let $C(M)$ be the set of all functions $\Omega(z)$ analytic inside the upper half-plane, whose values are contraction operators on $M : \|\Omega(z)\| \leq 1, \Im z > 0$. The following theorem gives a complete description of the set $S(A, M)$ of all finite positive operator measures $\sigma(\Delta)$ satisfying (1.16).

Theorem 1.7 *The formula*

$$\big(\mathcal{A}(z)\Omega(z) + \mathcal{B}(z)\big)\big(\mathcal{C}(z)\Omega(z) + \mathcal{D}(z)\big)^{-1} = \int_{-\infty}^{\infty} \frac{d\sigma(\lambda)}{\lambda - z}, \quad \Im z > 0, \tag{1.20}$$

determines a one-to-one correspondence

$$\sigma(\lambda) \leftrightarrow \Omega(z)$$

between the sets $S(A, M)$ and $C(M)$. Here the operator-functions $\mathcal{A}(z), \mathcal{B}(z), \mathcal{C}(z),$ $\mathcal{D}(z)$ are defined by the equality

$$\begin{pmatrix} \mathcal{A}(z) & \mathcal{B}(z) \\ \mathcal{C}(z) & \mathcal{D}(z) \end{pmatrix} = \begin{pmatrix} W_{11}(z) & W_{12}(z) \\ W_{21}(z) & W_{22}(z) \end{pmatrix} \begin{pmatrix} I_M & I_M \\ iI_M & -iI_M \end{pmatrix}. \tag{1.21}$$

It follows from this assertion that for an arbitrary $\Omega \in C(M)$ the left-hand side of (1.20) is meaningful for any complex number z. If $\|\Omega(z)\| < 1$ as $\Im z > 0$, then it can be written, by using (1.21), in the form

$$\big(W_{22}(z)\tau(z) + W_{21}(z)\big)\big(W_{12}(z)\tau(z) + W_{11}(z)\big)^{-1},$$

where

$$\tau(z) = i\big(I_M + \Omega(z)\big)\big(I_M - \Omega(z)\big)^{-1}. \tag{1.22}$$

Theorem 1.7 is a far-reaching generalization of Theorem 2.7.2 concerning the case of an entire operator with deficiency index $(1,1)$. Indeed, in the latter $\mathcal{P}(z)f = f_u(z)u$, where $f_u(z)$ is an entire scalar function, u is the corresponding entire gauge. If $\{d_k\}_{k=0}^\infty$ is an orthonormal basis in $\mathfrak{H}$, then

$$\mathcal{D}_k(z) = \big(\mathcal{P}(z)d_k, u\big) = \big(d_k, \mathcal{P}^*(z)u\big),$$
$$\mathcal{E}_k(z) = \big(R_z(d_k - \mathcal{D}_k(z)u), u\big) = \big(d_k, \mathcal{Q}^*(z)u\big),$$

and

$$\big(\mathcal{P}(z)\mathcal{P}^*(\zeta)u, u\big) = \sum_{k=0}^\infty \mathcal{D}_k(z)\overline{\mathcal{D}_k(\zeta)},$$

$$\big(\mathcal{Q}(z)\mathcal{Q}^*(\zeta)u, u\big) = \sum_{k=0}^\infty \mathcal{E}_k(z)\overline{\mathcal{E}_k(\zeta)},$$

$$\big(\mathcal{P}(z)\mathcal{Q}^*(\zeta)u, u\big) = \sum_{k=0}^\infty \mathcal{D}_k(z)\overline{\mathcal{E}_k(\zeta)},$$

$$\big(\mathcal{Q}(z)\mathcal{P}^*(\zeta)u, u\big) = \sum_{k=0}^\infty \mathcal{E}_k(z)\overline{\mathcal{D}_k(\zeta)}.$$

If we put

$$U(z) = \begin{pmatrix} p_1(z) & p_0(z) \\ q_1(z) & q_0(z) \end{pmatrix},$$

where

$$p_1(z) = \big(W_{22}(z)u, u\big) = 1 + z \sum_{k=0}^{\infty} \mathcal{E}_k(z)\overline{\mathcal{D}_k(0)},$$

$$p_0(z) = \big(W_{21}(z)u, u\big) = z \sum_{k=0}^{\infty} \mathcal{E}_k(z)\overline{\mathcal{E}_k(0)},$$

$$q_1(z) = \big(W_{12}(z)u, u\big) = -z \sum_{k=0}^{\infty} \mathcal{D}_k(z)\overline{\mathcal{D}_k(0)},$$

$$q_0(z) = \big(W_{11}(z)u, u\big) = 1 - z \sum_{k=0}^{\infty} \mathcal{D}_k(z)\overline{\mathcal{E}_k(0)},$$

we shall obtain Theorem 2.7.2. The relation (1.18) may be regarded as a generalization of the well-known scalar Christoffel identities

$$q_0(z)\overline{q_1(\zeta)} - q_1(z)\overline{q_0(\zeta)} = (z - \bar{\zeta}) \sum_{k=0}^{\infty} \mathcal{D}_k(z)\overline{\mathcal{D}_k(\zeta)},$$

$$p_0(z)\overline{p_1(\zeta)} - p_1(z)\overline{p_0(\zeta)} = (z - \bar{\zeta}) \sum_{k=0}^{\infty} \mathcal{E}_k(z)\overline{\mathcal{E}_k(\zeta)},$$

$$q_0(z)\overline{p_1(\zeta)} - q_1(z)\overline{p_0(\zeta)} = 1 - (z - \bar{\zeta}) \sum_{k=0}^{\infty} \mathcal{D}_k(z)\overline{\mathcal{E}_k(\zeta)}$$

(see the formulas (2.7.1)). The corresponding identities used in the classical moment problem and its various matrix and operator generalizations as well as in the continuation problem for Hermitian-positive scalar, matrix and operator functions, etc. are special cases of (1.18).

1.4 Let A again be a simple closed Hermitian operator with dense domain in $\mathfrak{H}$ whose defect numbers (finite or infinite) are equal. As has been done in section 2.9, construct the chain

$$\mathfrak{H}_+ \subseteq \mathfrak{H} \subseteq \mathfrak{H}_-\,,$$

where $\mathfrak{H}_+ = \mathcal{D}(A^*)$ with the graph-norm of A^*. The operator A^* continuously maps the space $\mathfrak{H}_+$ into $\mathfrak{H}$. Therefore the adjoint $\widehat{A}$ of the operator A^* defined as

$$\big(A^* f, g\big) = \big(f, \widehat{A}g\big), \quad f \in \mathfrak{H}_+, \quad g \in \mathfrak{H},$$

acts continuously from $\mathfrak{H}$ into $\mathfrak{H}_-$. It is an extension of the operator A, i.e.

$$\widehat{A}f = Af, \quad f \in \mathcal{D}(A)\,.$$

One can easily show that the linear manifold

$$\widehat{\mathfrak{M}}_z = (\widehat{A} - zI)\mathfrak{H}$$

is closed in the space $\mathfrak{H}_-$ for any $z \in \mathbb{C}^1 : \Im z \neq 0$ and

$$\widehat{\mathfrak{M}}_z = \overline{\mathfrak{M}_z}$$

(the closure is taken in the $\mathfrak{H}_-$-topology).

Fix a subspace $M \subset \mathfrak{H}_-$ (the so-called generalized gauge). A point $z \in \mathbb{C}^1$ is called M-regular for the operator A if it is its point of regular type and

$$\mathfrak{H}_- = M \dotplus \widehat{\mathfrak{M}}_z .$$

As in the case of an ordinary gauge, the set $\widehat{\rho}(A, M)$ of all M-regular points of A is open. We define on $\widehat{\rho}(A, M)$ the operator-function $\mathcal{P}(z)$ whose values are the projectors from $\mathfrak{H}_-$ onto M parallel to $\widehat{\mathfrak{M}}_z$. $\mathcal{P}(z)$ is called the function of the first kind. The function $T(z)$ of the second kind is defined as

$$T(z) = \left(\widehat{A} - zI\right)^{-1}\left(I - \mathcal{P}(z)\right) .$$

Its values belong to $L[\mathfrak{H}_-, \mathfrak{H}]$. In the same manner as for an ordinary gauge, one can prove that the operator-functions $\mathcal{P}(z)$ and $T(z)$ are analytic on $\widehat{\rho}(A, M)$ and possess the properties $1° - 7°$ of subsection 1.2.

Since a generalized resolvent R_z of the operator A acts continuously from $\mathfrak{H}$ into $\mathfrak{H}_+$, the operator $\widehat{R}_z$ which is defined by the relation

$$\left(R_z f, g\right) = \left(f, \widehat{R}_z g\right), \quad f \in \mathfrak{H}, \quad g \in \mathfrak{H}_-,$$

transforms $\mathfrak{H}_-$ into $\mathfrak{H}$ continuously and is an extension of $R_{\bar{z}}$:

$$\widehat{R}_z f = R_{\bar{z}} f, \quad f \in \mathfrak{H}.$$

We call the operator-function $\widehat{R}_z$ the extended generalized resolvent of A. When the resolvent R_z is orthogonal (canonical), its extension $\widehat{R}_z$ is called orthogonal, too.

Let E_λ be a spectral function of the operator A. Then for any finite interval $\Delta \subset \mathbb{R}^1$ the operator E_Δ acts continuously from $\mathfrak{H}$ into $\mathfrak{H}_+$. The operator $\widehat{E}_\Delta$ given as

$$\left(E_\Delta f, g\right) = \left(f, \widehat{E}_\Delta g\right), \quad f \in \mathfrak{H}, \quad g \in \mathfrak{H}_-,$$

belongs to the space $L[\mathfrak{H}_-, \mathfrak{H}]$. Taking into account the equality

$$\widehat{E}_\Delta^2 g = E_\Delta \widehat{E}_\Delta g, \quad g \in \mathfrak{H}_-,$$

we conclude that $\widehat{E}_\Delta \in L[\mathfrak{H}_-, \mathfrak{H}_+]$ and $\widehat{E}_\Delta$ is an extension of E_Δ onto $\mathfrak{H}_-$. We shall call the operator-function

$$\widehat{E}_\lambda = \begin{cases} \widehat{E}_{[0,\lambda)} & \text{as } \lambda > 0 \\ \widehat{E}_{[\lambda,0)} & \text{as } \lambda < 0 \end{cases}$$

the extended spectral function of the operator A. For any $f, g \in \mathfrak{H}$ the scalar function $(\widehat{E}_\lambda f, g) = (f, E_\lambda g)$ belongs to $\mathfrak{S}^0$. If $f \in \mathfrak{H}, g \in \mathfrak{H}_-$, then $(\widehat{E}_\lambda f, g) \in \mathfrak{S}^1$ and $(\widehat{E}_\lambda f, g) \in \mathfrak{S}^2$ when $f, g \in \mathfrak{H}_-$.

The analogue of Theorem 1.4 holds in the case of a generalized gauge, namely: if for a vector $f_0 \in \mathfrak{H}_-$ the vector-function $\mathcal{P}(z) f_0$ admits an analytic continuation to a closed interval Δ_0 of the real axis, then so does the vector-function $T(z) f_0$ and

$$\widehat{E}_{\Delta_0} f_0 = \int_{\Delta_0} d\widehat{E}_t \mathcal{P}(t) f_0 \,,$$

where $\widehat{E}_\lambda$ is an extended spectral function of the operator A. The proof is similar to that of Theorem 1.4.

Define the operator measure $\sigma(\Delta)$ on M by the equality

$$(\sigma(\Delta) f, g)_- = (\widehat{E}_\Delta f, g), \quad f, g \in M, \quad \Delta \subset \mathbb{R}^1 \text{ is a finite interval.} \tag{1.23}$$

The function σ belongs to the class $\mathfrak{S}^2(M, M)$. The assertion analogous to Theorem 1.5 is valid. It consists of the following: if for some $f, g \in \mathfrak{H}_-$ the vector-functions $\mathcal{P}(\lambda) f$ and $\mathcal{P}(\lambda) g$ are analytically continuable to an interval $\Delta_0 = [a, b] \subset \mathbb{R}^1$, then

$$(\widehat{E}_{\Delta_0} f, g) = \int_{\Delta_0} (d\sigma(t) \mathcal{P}(t) f, \mathcal{P}(t) g)_- , \tag{1.24}$$

which is a generalization of (1.13). The operator A is called M-entire if the set $\widehat{\rho}(A, M)$ of its all M-regular points coincides with the whole complex plane.

In the case when the operator A is M-entire the functions $\mathcal{P}(z)$ and $T(z)$ taking their values in $L[\mathfrak{H}_-]$ are entire. We assign to each vector $f \in \mathfrak{H}_-$ the entire function

$$f_M(z) = \mathcal{P}(z) f$$

with values in M. The mapping

$$f \mapsto f_M(z), \quad f \in \mathfrak{H}_-,$$

transforms the space $\mathfrak{H}_-$ into a certain space of entire vector-functions whose values belong to M, and the equality

$$(f, g) = \int_{-\infty}^{\infty} (d\sigma(t) \mathcal{P}(t) f, \mathcal{P}(t) g)_- \tag{1.25}$$

is fulfilled for any $f, g \in \mathfrak{H}$, where the operator measure $\sigma(\Delta)$ is determined from the relation (1.23). It turns out that the converse is valid, too. If for a non-negative measure $\sigma(\Delta)$ with values in $L[M]$ the equality (1.25) holds for any $f, g \in \mathfrak{H}$, then an extended spectral function $\widehat{E}_\lambda$ of the operator A can be found so that the equations (1.23) and (1.24) are satisfied. As in the case of an ordinary gauge, it is possible to give the analytical description of the set of all measures $\sigma(\Delta)$ possessing the property (1.25). Just note that, generally, the expression $(\widehat{R}_\lambda f, f)$, where $\widehat{R}_\lambda$ is an extended generalized resolvent of A, may have no sense as $f \in \mathfrak{H}_-$. In a way analogous to that used in subsection 2.9.2 one can verify that there exists an operator H self-adjoint on $\mathfrak{H}$, which maps $\mathfrak{H}_-$ into $\mathfrak{H}$ continuously and $\mathfrak{H}$ into $\mathfrak{H}_+$, such that the operator

$$\widehat{R}_{\lambda,H} = \widehat{R}_\lambda - H$$

is continuous from $\mathfrak{H}_-$ into $\mathfrak{H}_+$ and $(\widehat{R}_{\lambda,H} f, f)$ is an R-function for any $f \in \mathfrak{H}_-$. So,

$$\widehat{R}_{\lambda,H} = Q_H + \int_{-\infty}^{\infty} \left(\frac{1}{t-\lambda} - \frac{t}{t^2+1} \right) d\widehat{E}_t \, ;$$

here the operator $Q_H \in L[\mathfrak{H}_-, \mathfrak{H}]$ is self-adjoint on $\mathfrak{H}$, E_t is the spectral function of the operator A associated with the generalized resolvent R_λ. As has been mentioned in subsection 2.9.2, we can put

$$H = \frac{1}{2} \left(\widehat{\overset{\circ}{R}}_{\lambda_0} + \widehat{\overset{\circ}{R}}_{\overline{\lambda_0}} \right),$$

where $\widehat{\overset{\circ}{R}}_{\lambda_0}$, $\Im\lambda_0 \neq 0$, is an orthogonal extended resolvent of the operator A. A certain matrix-function

$$W(z) = \begin{pmatrix} W_{11}(z) & W_{12}(z) \\ W_{21}(z) & W_{22}(z) \end{pmatrix}$$

with values in $L[M \oplus M]$ corresponds to the operator A entire with respect to the generalized gauge M so that the relation (1.18) holds, where

$$Q(z) = P_M T_H(z), \quad z \in \mathbb{C}^1, \quad T_H(z) = T(z) - HP(z)$$

(P_M is the orthoprojector from $\mathfrak{H}_-$ onto M). The function $W(z)$ is uniquely determined up to a constant $\mathcal{J}$-unitary factor $\mathcal{E}$. Therefore, it can be chosen so that $W(0) = I$. Then the description of all operator-functions $\sigma(\lambda)$ with the property (1.25) is given by the formula

$$\left(W_{22}(z)\tau(z) + W_{21}(z) \right) \left(W_{12}(z)\tau(z) + W_{11}(z) \right)^{-1}$$

$$= C + \int_{-\infty}^{\infty} \left(\frac{1}{t-z} - \frac{t}{t^2+1} \right) d\sigma(t), \tag{1.26}$$

where $\tau(z)$ is the operator-function determined by the equality (1.22) in which $\Omega(z)$ goes through the set of all operator-functions with values in $L[M]$ analytic in the half-plane $\Im z > 0$, and such that $\|\Omega(z)\| \leq 1, C \in L[M]$.

Appendix 2

Entire Operators Being Represented by Differential Operators

It is shown here that the minimal operator generated by a second order operator-differential expression of hyperbolic type is entire with respect to a generalized gauge whose defect numbers are infinite. The effective formulas are given for the functions of the first and the second kind. The set of all spectral functions of the minimal operator is described. Note that this case covers a lot of partial differential equations of hyperbolic type since the coefficients of the considered expression are, generally, unbounded.

2.1   We consider a differential expression of the form

$$l\,[y]\,(t) = y''(t) + \mathcal{A}y(t) - q(t)y(t), \quad t \in [0, b], \quad 0 < b < \infty, \tag{2.1}$$

where $\mathcal{A}$ is a lower semibounded self-adjoint operator on a separable Hilbert space $\mathcal{H}$ (without loss of generality we can suppose $\mathcal{A} \geq 0$), $q(t) = q^*(t)$ is a strongly continuous function on $[0, b]$ with values in the set $L[\mathcal{H}]$ of all bounded operators on $\mathcal{H}$. Moreover, the operator-function $(\mathcal{A} + I)q(t)(\mathcal{A} + I)^{-1}$ is assumed to be strongly continuous on $[0, b]$. The latter implies also the strong continuity on $[0, b]$ of the function $(\mathcal{A} + I)^{\frac{1}{2}}q(t)(\mathcal{A} + I)^{-\frac{1}{2}}$. In the case where $\mathcal{H} = \mathbb{C}^1$, $\mathcal{A} = 0$, $q(t)$ is a real scalar function continuous on $[0, b]$, the expression (2.1) becomes the ordinary self-adjoint Sturm-Liouville one. If $\mathcal{H} = L_2(\mathbb{R}^1)$, $y(t) = u(t, \cdot)$, $\mathcal{A} = -\frac{d^2}{dx^2}$, $q(t)y(t) = q(t, x)u(t, x)$, $q(t, x)$ is a real continuous function bounded on $[0, b] \times \mathbb{R}^1$, then $l\,[y]$ coincides with the partial differential expression

$$\frac{\partial^2 u(t, x)}{\partial t^2} - \frac{\partial^2 u(t, x)}{\partial x^2} - q(t, x)u(t, x)$$

of hyperbolic type.

The minimal operator L_0 generated on the space

$$\mathfrak{H} = L_2(\mathcal{H}, [0, b])$$

by the expression (2.1) and the boundary condition

$$y'(0) = 0 \tag{2.2}$$

is defined as follows. Let $\mathcal{D}'_0$ be the set of all elements of the form

$$y(t) = \sum_{k=1}^{m} \varphi_k(t) f_k,$$

where $f_k \in \mathcal{D}(\mathcal{A})$, $\varphi_k(t)$ is a scalar twice continuously differentiable function on $[0, b]$ vanishing in some neighbourhood of the point b and satisfying the condition $\varphi_k'(0) = 0$. Put

$$L_0' y = l\,[y], \quad \mathcal{D}(L_0') = \mathcal{D}_0'.$$

The integration by parts formula

$$\int_\alpha^\beta \left(l\,[y]\,(t), z(t)\right)_\mathcal{H} dt - \int_\alpha^\beta \left(y(t), l\,[z]\,(t)\right)_\mathcal{H} dt \tag{2.3}$$

$$= \left(y'(t), z(t)\right)_\mathcal{H} - \left(y(t), z'(t)\right)_\mathcal{H} \Big|_\alpha^\beta, \quad 0 \le \alpha < \beta \le b$$

valid for functions twice continuously differentiable on $[0, b]$ with values in $\mathcal{D}(\mathcal{A})$ shows that the operator L_0' is Hermitian in $\mathfrak{H}$. The operator $L_0 = \overline{L_0'}$ is called minimal. Its adjoint L_0^* is called the maximal one.

Denote by $\mathcal{H}_j$, $-\infty < j < \infty$, the Hilbert scale of the spaces constructed from the operator $\mathcal{A}$:

$$\mathcal{H}_j = \mathcal{D}((\mathcal{A} + I)^j), \quad (f, g)_{\mathcal{H}_j} = \left((\mathcal{A} + I)^j f, (\mathcal{A} + I)^j g\right)_\mathcal{H}, \quad j \ge 0,$$

$\mathcal{H}_{-j}$ is the space with negative norm in the sense of Berezansky [1], or $\mathcal{H}_{-j}$, $j > 0$, is the completion of $\mathcal{H}$ in the norm

$$\|f\|_{\mathcal{H}_{-j}} = \left\|(\mathcal{A} + I)^{-j} f\right\|_\mathcal{H},$$

which is the same. As in Appendix 1, $\hat{\mathcal{A}}$ denotes the adjoint of the operator $\mathcal{A}$ considered as the operator from $\mathcal{H}_1$ into $\mathcal{H}$. The operator $\hat{\mathcal{A}}$ acts from the space $\mathcal{H}$ into $\mathcal{H}_{-1}$.

Definition 2.1 *By a solution of the equation*

$$l\,[y]\,(t) = \lambda y(t), \quad \lambda \in \mathbb{C}^1, \quad t \in [0, b], \tag{2.4}$$

we mean a twice differentiable in $\mathcal{H}$ vector-function $y(t) : [0, b] \to \mathcal{D}(\mathcal{A})$ satisfying (2.4). We call a function $y(t) : [0, b] \to \mathcal{H}$ a weak solution of the equation (2.4) if this function is weakly differentiable in $\mathcal{H}_{-\frac{1}{2}}$ and twice weakly differentiable in $\mathcal{H}_{-1}$ (i.e. the scalar function $(y(t), f)_\mathcal{H}$ is twice differentiable for any $f \in \mathcal{H}_1$), and

$$\frac{d^2}{dt^2}\left(y(t), f\right)_\mathcal{H} + \left(y(t), \mathcal{A}f\right)_\mathcal{H} - \left(y(t), q(t)f\right)_\mathcal{H} = \lambda\left(y(t), f\right)_\mathcal{H}, \quad f \in \mathcal{H}_1.$$

Set

$$\hat{l}\,[y]\,(t) = y''(t) + \hat{\mathcal{A}}y(t) - q(t)y(t).$$

The fact that a vector-function $y(t)$ is a weak solution of the equation (2.4) is equivalent to the weak differentiability of $y(t)$ in $\mathcal{H}_{-1}$ and the identity

$$\hat{l}\,[y]\,(t) = \lambda y(t), \quad t \in [0, b]. \tag{2.5}$$

If

a) a vector-function $y(t) : [0, b] \to \mathcal{H}$ is twice continuously weakly differentiable in $\mathcal{H}_{-1}$, and a vector-function $z(t) : [0, b] \mapsto \mathcal{H}_1$ is and twice differentiable in $\mathcal{H}_1$,

or

b) a vector-function $y(t) : [0, b] \to \mathcal{H}$ is weakly differentiable in $\mathcal{H}_{-\frac{1}{2}}$ and twice continuously weakly differentiable in $\mathcal{H}_{-1}$, and a vector-function $z(t) : [0, b] \to \mathcal{H}_1$ is continuously differentiable in $\mathcal{H}_1$ and twice continuously differentiable in $\mathcal{H}_{\frac{1}{2}}$,

then the formula

$$\int_\alpha^\beta \left(\hat{l}\,[y]\,(t), z(t) \right)_{\mathcal{H}} dt - \int_\alpha^\beta \left(y(t), \hat{l}\,[z](t) \right)_{\mathcal{H}} dt \tag{2.6}$$

$$= \left(y'(t), z(t) \right)_{\mathcal{H}} - \left(y(t), z'(t) \right)_{\mathcal{H}} \Big|_\alpha^\beta, \quad 0 \le \alpha < \beta \le b,$$

holds.

Using the method of successive approximation, it is not hard to show that the integral equations

$$\omega_1(t, x, \lambda) = \cos \sqrt{A - \lambda I}\,(t - x) + \int_x^t \frac{\sin \sqrt{A - \lambda I}\,(t - \xi)}{\sqrt{A - \lambda I}}\, q(\xi) \omega_1(\xi, x, \lambda)\, d\xi$$

and

$$\omega_2(t, x, \lambda) = \frac{\sin \sqrt{A - \lambda I}\,(t - x)}{\sqrt{A - \lambda I}} + \int_x^t \frac{\sin \sqrt{A - \lambda I}\,(t - \xi)}{\sqrt{A - \lambda I}}\, q(\xi) \omega_2(\xi, x, \lambda)\, d\xi$$

$$x \in [0, b], \quad t \in [0, b],$$

are solvable, the first equation in the class of strongly continuous operator-functions, and the second one in the class of operator-functions continuous in the uniform operator topology jointly in the variables t and x. For fixed $t, x \in [0, b]$ the solutions $\omega_1(t, x, \lambda)$ and $\omega_2(t, x, \lambda)$ are entire operator-functions in λ.

The operational calculus for self-adjoint operators allows us to conclude that for $f \in \mathcal{H}_1$, $g \in \mathcal{H}_{\frac{1}{2}}$ the vector-functions $\omega_1(t, x, \lambda)f$ and $\omega_2(t, x, \lambda)g$ are the solutions in t of the equation (2.4), continuous in $\mathcal{H}_1$, continuously differentiable in $\mathcal{H}_{\frac{1}{2}}$ and satisfying the conditions

$$\omega_1(x, x, \lambda)f = f, \quad \omega_1'(x, x, \lambda)f = 0; \tag{2.7}$$

$$\omega_2(x, x, \lambda)g = 0, \quad \omega_2'(x, x, \lambda)g = g. \tag{2.8}$$

If $g \in \mathcal{H}_1$, then the vector-function $\omega_2(t, x, \lambda)g$ is continuously differentiable in $\mathcal{H}_1$. For $f \in \mathcal{H}, g \in \mathcal{H}_{-\frac{1}{2}}$, the functions $\omega_1(t, x, \lambda)f$ and $\omega_2(t, x, \lambda)g$ are the weak solutions in t of this equation which satisfy the conditions (2.7), (2.8).

Put

$$\omega_1(t, \lambda) = \omega_1(t, 0, \lambda),$$

$$\omega_2(t, \lambda) = \omega_2(t, 0, \lambda).$$

Lemma 2.1 *The Cauchy problem*

$$y(0) = f, \quad y'(0) = g, \quad f \in \mathcal{H}, \quad g \in \mathcal{H}_{-\frac{1}{2}},$$

for the equation (2.4) *has a unique weak solution. This solution is of the form*

$$y(t, \lambda) = \omega_1(t, \lambda)f + \omega_2(t, \lambda)g. \tag{2.9}$$

Proof. If $f \in \mathcal{H}$, $g \in \mathcal{H}_{-\frac{1}{2}}$, then the vector-functions $\omega_1(t, \lambda)f$ and $\omega_2(t, \lambda)g$ are the weak solutions of the equation (2.4), satisfying the conditions (2.7),(2.8) respectively when $x = 0$. So, a vector-function $y(t, \lambda)$ of the form (2.9) is a weak solution of this equation such that

$$y(0, \lambda) = f, \quad y'(0, \lambda) = g.$$

Let us show now that if $w(t, \lambda)$ is a weak solution of (2.4) satisfying the conditions

$$w(0, \lambda) = 0, \quad w'(0, \lambda) = 0,$$

then $w(t, \lambda) \equiv 0$ on the interval $[0, b]$. Indeed, let x be an arbitrary fixed point from $[0, b]$. Multiplying scalarly by $w(t, \lambda)$ the equality

$$\frac{d^2}{dt^2}w_2(t, x, \overline{\lambda})h + \mathcal{A}w_2(t, x, \overline{\lambda})h - q(t)w_2(t, x, \overline{\lambda})h = \overline{\lambda}w_2(t, x, \overline{\lambda})h, \quad h \in \mathcal{H}_1,$$

and then integrating, we obtain

$$\int_0^x \left(w(t, \lambda), \omega_2''(t, x, \overline{\lambda})h\right)_\mathcal{H} dt + \int_0^x \left(w(t, \lambda), \mathcal{A}w_2(t, x, \overline{\lambda})h\right)_\mathcal{H} dt$$

$$- \int_0^x \left(w(t, \lambda), q(t)w_2(t, x, \overline{\lambda})h\right)_\mathcal{H} dt = \lambda \int_0^x \left(w(t, \lambda), w_2(t, x, \overline{\lambda})h\right)_\mathcal{H} dt.$$

As was mentioned above, the functions $w_2(t, x, \overline{\lambda})h$, $h \in \mathcal{H}_1$, and $w(t, \lambda)$ satisfy the condition b). Therefore,

$$\int_0^x \left(w(t, \lambda), l\left[w_2(\cdot, x, \overline{\lambda})h \right](t) - \overline{\lambda} w_2(t, x, \overline{\lambda})h \right)_{\mathcal{H}} dt$$

$$- \int_0^x \left(\hat{l}\left[w(\cdot, \lambda) \right](t) - \lambda w(t, \lambda), w_2(t, x, \overline{\lambda})h \right)_{\mathcal{H}} dt$$

$$= \left(w(t, \lambda), w_2'(t, x, \overline{\lambda})h \right)_{\mathcal{H}} - \left(w'(t, \lambda), w_2(t, x, \overline{\lambda})h \right)_{\mathcal{H}} \Big|_0^x .$$

Since

$$\hat{l}\left[w(\cdot, \lambda) \right](t) - \lambda w(t, \lambda) = 0, \qquad l\left[w_2(\cdot, x, \overline{\lambda})h \right](t) - \overline{\lambda} w_2(t, x, \overline{\lambda})h = 0,$$
$$w(0, \lambda) = 0, \quad w'(0, \lambda) = 0, \qquad \text{and} \qquad w_2(x, x, \overline{\lambda})h = 0, \quad w_2'(x, x, \overline{\lambda})h = h,$$

$\left(w(x, \lambda), h \right)_{\mathcal{H}} = 0$ for any $h \in \mathcal{H}_1$. Taking into account the equality $\overline{\mathcal{H}_1} = \mathcal{H}$ and arbitrariness of $x \in [0, b]$, we conclude that $w(t, \lambda) \equiv 0$. □

Note that under the assumption $f \in \mathcal{H}_1$, $g \in \mathcal{H}_{\frac{1}{2}}$ the vector-function $y(t, \lambda)$ of the form (2.9) is a solution of the Cauchy problem for the equation (2.4). It follows that in the case of a bounded $\mathcal{A}$ every weak solution of the equation (2.4) is a solution of this equation, because $\mathcal{H}_j = \mathcal{H}$ for any $j \in \mathbb{R}^1$ in this situation.

The substitution of

$$y(t) = w_i(t, \lambda)f, \quad z(t) = w_j(t, \overline{\lambda})g, \quad i, j = 1, 2, \quad f, g \in \mathcal{D}(\mathcal{A}),$$
$$\alpha = 0, \ \beta = t \in [0, b],$$

in the formula (2.6) yields $(i = 1, 2)$

$$w_i'^*(t, \overline{\lambda}) w_i(t, \lambda) - w_i^*(t, \overline{\lambda}) w_i'(t, \lambda) = 0,$$
$$w_1'^*(t, \overline{\lambda}) w_2(t, \lambda) - w_1^*(t, \overline{\lambda}) w_2'(t, \lambda) = -I, \qquad (2.10)$$
$$w_2'^*(t, \overline{\lambda}) w_1(t, \lambda) - w_2^*(t, \overline{\lambda}) w_1'(t, \lambda) = I.$$

2.2   The domains of the minimal and maximal operators are investigated in this subsection.

Lemma 2.2 *The domain $\mathcal{D}(L_0^*)$ of the maximal operator L_0^* generated by the expression (2.1) and the boundary condition (2.2) consists of those and only those vector-functions $y(t) \in \mathfrak{H} = L_2\left(\mathcal{H}, [0, b]\right)$ which can be represented as*

$$y(t) = w_1(t, \lambda)f + \int_0^t w_2(t, x, \lambda)h(x)\, dx, \qquad (2.11)$$

where $h = (L_0^* - \lambda I)y$, $f \in \mathcal{H}$, λ is any complex number.

Proof. Let $y \in \mathcal{D}(L_0^*)$. Then

$$(L_0 z, y) = (z, h) + \lambda (z, y), \quad z \in \mathcal{D}(L_0). \tag{2.12}$$

Suppose first $h = 0$. Since a vector-function of the form $z(t) = \varphi(t) f$, where $f \in \mathcal{D}(\mathcal{A}), \varphi(t)$ is a scalar infinitely differentiable function on $[0, b]$ vanishing in some neighbourhoods of the ends of the interval, belongs to $\mathcal{D}_0'$,

$$\int_0^b \varphi''(t) \left(f, y(t)\right)_{\mathcal{H}} dt = \int_0^b \varphi(t) F(t) \, dt \, ; \tag{2.13}$$

here

$$F(t) = \left(q(t)f - (\mathcal{A} - \lambda I)f, y(t)\right)_{\mathcal{H}}.$$

Set

$$w(t) = \int_0^t (t - s)\left(q(s)y(s) - (\widehat{\mathcal{A}} - \overline{\lambda}I)y(s)\right) ds \, .$$

The function $w(t)$ takes its values in $\mathcal{H}_{-1}$. Its first derivative is absolutely continuous in $\mathcal{H}_{-1}$, the second one is square integrable in the space $\mathcal{H}_{-1}$. Passing to the limit and integrating by parts, we obtain the equality

$$\int_0^b \varphi''(t) \left(f, y_1(t)\right)_{\mathcal{H}} dt = 0 \tag{2.14}$$

valid for each $\varphi(t) \in \overset{\circ}{W}_2^2(0, b)$ where

$$y_1(t) = y(t) - w(t) - g_1 - g_2 t \, ,$$

$g_1, g_2 \in \mathcal{H}_{-1}$ are arbitrary, $\overset{\circ}{W}_2^2(0, b)$ is the set of all scalar differentiable functions $\varphi(t)$ on $[0, b]$ whose first derivatives are absolutely continuous, the second derivatives belong to $L_2(0, b)$, and such that $\varphi(0) = \varphi'(0) = \varphi(b) = \varphi'(b) = 0$. Choosing the vectors g_1 and g_2 so that the function

$$\psi(t) = \int_0^t (t - s) y_1(s) \, ds$$

satisfies the conditions $\psi(b) = \psi'(b) = 0$ (g_1 and g_2 are uniquely determined by these conditions) and substituting $\varphi(t) = \left(\psi(t), f\right)_{\mathcal{H}}$ into (2.14), we get

$$\int_0^b \left|\left(y_1(t), f\right)_{\mathcal{H}}\right|^2 dt = 0$$

for any $f \in \mathcal{H}_1$, whence $y_1(t) = 0$ almost everywhere on $[0, b]$. Integration by parts in (2.13) yields (2.5). Thus the equality

$$\hat{l}\,[y]\,(t) = \lambda y(t)$$

is fulfilled almost everywhere.

If we take in (2.12) $h = 0$, $z(t) = \varphi(t)f$, where $f \in \mathcal{D}(\mathcal{A})$ and $\varphi(t)$ is an infinitely differentiable scalar function vanishing in a neighbourhood of the point b such that $\varphi(0) = 1, \varphi'(0) = 0$, we shall find, using the relation (2.3), that

$$\left(f, y'(0)\right)_{\mathcal{H}} = 0, \quad f \in \mathcal{H}_1,$$

hence $y'(0) = 0$.

Thus, in the case of $h = 0$ a vector-function $y(t) \in \mathcal{D}(L_0^*)$ has the following properties: $y'(t)$ is absolutely continuous in $\mathcal{H}_{-1}$, $y'' \in L_2(\mathcal{H}_{-1}, [0, b])$, and $y'(0) = 0$. In a way analogous to that used in the proof of Lemma 2.1, we obtain

$$y(t) = \omega_1(t, \lambda)f, \quad f \in \mathcal{H}.$$

Now let $h \neq 0$. It is easily verified that the function

$$v(t) = \int\limits_0^t \omega_2(t, x, \lambda)h(x)\,dx$$

is continuously differentiable in $\mathcal{H}$, its derivative is absolutely continuous in $\mathcal{H}_{-\frac{1}{2}}$, and $v'' \in L_2(\mathcal{H}_{-\frac{1}{2}}, [0, b])$. The function $v(t)$ satisfies the equation

$$\hat{l}\,[v] - \lambda v = h \quad \text{and the conditions} \quad v(0) = v'(0) = 0.$$

In view of (2.6) and (2.12),

$$\left((L_0 - \lambda I)z, u\right) = 0, \quad z \in \mathcal{D}(L_0),$$

where $u(t) = y(t) - v(t)$ satisfies the condition $u'(0) = 0$. This implies the representation (2.11).

The converse follows immediately from the formula (2.6). □

Corollary 2.1 *A vector-function $y(t)$ belongs to the defect subspace $\mathfrak{N}_\lambda$ of the minimal operator L_0 if and only if it admits the representation*

$$y(t) = \omega_1(t, \lambda)f, \quad f \in \mathcal{H}.$$

Since the operator L_0^* differs from the maximal operator generated by the expression (2.1) with $q(t) \equiv 0$ and the condition (2.2) by a certain bounded operator, the following assertion is true.

Corollary 2.2 *In order that a vector-function $y(t)$ belong to $\mathcal{D}(L_0^*)$, it is necessary and sufficient that*

$$y(t) = \cos \sqrt{\mathcal{A} - \lambda I}\, t \, f + \int_0^t \frac{\sin \sqrt{\mathcal{A} - \lambda I}\,(t - x)}{\sqrt{\mathcal{A} - \lambda I}}\, h(x)\, dx\,, \qquad (2.15)$$

where $f \in \mathcal{H}, h \in L_2(\mathcal{H}, [0, b])$.

The next theorem follows from the representation (2.15) and the integration by parts formula (2.6).

Theorem 2.1 *A vector-function $y(t)$, $t \in [0, b]$, belongs to the domain $\mathcal{D}(L_0^*)$ of the maximal operator L_0^* generated on the space $\mathfrak{H} = L_2(\mathcal{H}, [0, b])$ by the expression (2.1) and the boundary condition (2.2) if and only if it satisfies the conditions:*

1) $y(t)$ *is continuous in $\mathcal{H}$;*

2) $y'(t)$ *is continuous in $\mathcal{H}_{-\frac{1}{2}}$ and absolutely continuous in $\mathcal{H}_{-1}$;*

3) $\hat{l}\,[y] \in \mathfrak{H}$;

4) $y'(0) = 0.$

Moreover,
$$L_0^* y = \hat{l}\,[y]\,.$$

Denote by $\mathcal{D}'$ the set of all vector-functions $y(t)$, $t \in [0, b]$, continuous in $\mathcal{H}_1$, twice continuously differentiable in $\mathcal{H}$, and satisfying the condition $y'(0) = 0$, and by L' the restriction of L_0^* to $\mathcal{D}'$.

Corollary 2.3 *The operator L_0^* coincides with the closure in $\mathfrak{H}$ of the operator L'.*

Proof. Let $y \in \mathcal{D}(L_0^*)$ and $L_0^* y = h$. According to Lemma 2.2,

$$y(t) = \omega_1(t, 0) f + \int_0^t \omega_2(t, x, 0) h(x)\, dx\,, \qquad f \in \mathcal{H}\,.$$

Choose a sequence $f_n \in \mathcal{D}(\mathcal{A})$ convergent to f in $\mathcal{H}$ and a sequence $h_n(t)$ of continuously differentiable in $\mathcal{H}_1$ vector-functions convergent to h in $\mathfrak{H}$. Then the functions

$$y_n(t) = \omega_1(t, 0) f_n + \int_0^t \omega_2(t, x, 0) h_n(x)\, dx$$

belong to $\mathcal{D}', y_n \to y$, and $L' y_n = h_n \to h$ in $\mathfrak{H}$ as $n \to \infty$. $\qquad\square$

Theorem 2.2 *The domain $\mathcal{D}(L_0)$ of the minimal operator L_0 generated on $\mathfrak{H}$ by the expression (2.1) and the condition (2.2) consists of those and only those vector-functions $y(t) \in \mathcal{D}(L_0^*)$ which satisfy the conditions:*

1) *$y(t)$ is continuously differentiable in $\mathcal{H}$ on $[0,b]$;*

2) *$y(b) = y'(b) = 0$.*

Proof. Let $y \in \mathcal{D}(L_0)$. Since $L_0 = L_0^{**} = L'^*$, we have

$$\int_0^b \left(l\,[z]\,(t), y(t) \right)_{\mathcal{H}} dt = \int_0^b \left(z(t), h(t) \right)_{\mathcal{H}} dt, \quad h = L_0 y,$$

for an arbitrary $z \in \mathcal{D}(L')$. As $y \in \mathcal{D}(L_0^*)$, we can apply (see the case b)) the integration by parts formula (2.6) to the vector-functions $y(t)$ and $z(t) \in \mathcal{D}(L')$. This implies

$$\left(z'(b), y(b) \right)_{\mathcal{H}} - \left(z(b), y'(b) \right)_{\mathcal{H}} = 0.$$

Taking $z(t) = \psi(t)f$, where $f \in \mathcal{D}(A), \psi(t)$ is a scalar infinitely differentiable function such that $\psi(0) = \psi'(b) = 0, \psi(b) = 1$, we find that $\left(f, y'(b) \right)_{\mathcal{H}} = 0$, whence $y'(b) = 0$. If $\psi(t)$ is selected so that $\psi(0) = \psi'(0) = 0, \psi'(b) = 1$, we obtain $\left(f, y(b) \right)_{\mathcal{H}} = 0$, i.e. $y(b) = 0$. Thus, we have proved that if $y \in \mathcal{D}(L_0)$, then

$$y(b) = y'(b) = 0. \tag{2.16}$$

Substituting the representation (2.15), where $\lambda = -1$, for y, we arrive at the equalities

$$\cos \sqrt{A + I}\, b\, f + \frac{\sin \sqrt{A + I}\, b}{\sqrt{A + I}} f_1 - \frac{\cos \sqrt{A + I}\, b}{\sqrt{A + I}} f_2 = 0, \tag{2.17}$$

$$-\sqrt{\widehat{A} + I} \sin \sqrt{\widehat{A} + I}\, b\, f + \cos \sqrt{\widehat{A} + I}\, b\, f_1 + \sin \sqrt{\widehat{A} + I}\, b\, f_2 = 0, \tag{2.18}$$

where

$$f_1 = \int_0^b \cos \sqrt{A + I}\, t\, h(t)\, dt, \quad f_2 = \int_0^b \sin \sqrt{A + I}\, t\, h(t)\, dt,$$

$$h(t) = -y''(t) + \left(\widehat{A} + I \right) y(t) \in \mathfrak{H}.$$

Apply the operator $\left(\widehat{A} + I \right)^{-\frac{1}{2}}$ to both sides of (2.18). Adding the equality obtained multiplied by $-\sin \sqrt{A + I}\, b$ and the equality (2.17) multiplied by $\cos \sqrt{A + I}\, b$, we get

$$f - (A + I)^{-\frac{1}{2}} f_2 = 0$$

which shows that $f \in \mathcal{H}_{\frac{1}{2}}$. The continuous differentiability in $\mathcal{H}$ of the vector-function $y(t)$ follows from the representation (2.15).

The sufficiency is established immediately on the basis of Corollary 2.3 and the formula (2.3). $\qquad\square$

Corollary 2.4 *If $y \in \mathcal{D}(L_0)$ and $\chi(t)$ is a twice continuously differentiable scalar function such that $\chi'(0) = 0$, then $z(t) = \chi(t)y(t) \in \mathcal{D}(L_0)$.*

Proof. It is clear that the vector-function $z'(t)$ is continuous in $\mathcal{H}$ and absolutely continuous in $\mathcal{H}_{-1}$. Next,

$$\hat{l}\,[z]\,(t) = \chi''(t)y(t) + 2\chi'(t)y'(t) + \chi(t)\hat{l}\,[y]\,(t)\,.$$

By Theorems 2.1 and 2.2, $\hat{l}\,[z] \in L_2(\mathcal{H}, [0, b])$. Taking into account that $y(b) = y'(b) = y'(0) = 0$, one can check directly that $z(b) = z'(b) = z'(0) = 0$. $\qquad\square$

Corollary 2.5 *Every point $\lambda \in \mathbb{C}^1$ is a point of regular type for the minimal operator L_0.*

Proof. Let $y \in \mathcal{D}(L_0)$. According to Lemma 2.2,

$$y(t) = w_1(t, \lambda)f + \int_0^t w_2(t, x, \lambda)h(x)\,dx\,,$$

where $f \in \mathcal{H}$, $h = (L_0 - \lambda I)y$. The substitution of this expression for $y(t)$ in the boundary conditions (2.16) gives

$$w_1(b, \lambda)f + \int_0^b w_2(b, x, \lambda)h(x)\,dx = 0\,;$$

$$w_1'(b, \lambda)f + \int_0^b w_2'(b, x, \lambda)h(x)\,dx = 0\,.$$

Applying the operator $w_2'^*(b, \bar{\lambda})$ to the first identity and $w_2^*(b, \bar{\lambda})$ to the second and then subtracting, we get, due to (2.10),

$$f = \int_0^b K(x, \lambda)h(x)\,dx\,,$$

where

$$K(x, \lambda) = -w_2'^*(b, \bar{\lambda})w_2(b, x, \lambda) + w_2^*(b, \bar{\lambda})w_2'(b, x, \lambda)$$
$$= -w_2'^*(0, \bar{\lambda})w_2(0, x, \lambda) + w_2^*(0, \bar{\lambda})w_2'(0, x, \lambda) = -w_2(0, x, \lambda)\,.$$

Thus,

$$y(t) = -\int_0^b w_1(t, \lambda)w_2(0, x, \lambda)h(x)\,dx + \int_0^t w_2(t, x, \lambda)h(x)\,dx\,. \qquad (2.19)$$

Since the functions $w_1(t, \lambda)w_2(0, x, \lambda)$ and $w_2(t, x, \lambda)$ are continuous on $[0, b]$ in the uniform operator topology,

$$\|y\| \le c\,\|h\| = c\,\big\|(L_0 - \lambda I)y\big\|,$$

which is what had to be proved. □

2.3 Put

$$\mathcal{I}_\lambda = \int\limits_0^b w_1^*(t, \lambda)w_1(t, \lambda)\,dt.$$

Lemma 2.3 *The operator $\mathcal{I}_\lambda$ on $\mathcal{H}$ is invertible for any $\lambda \in \mathbb{C}^1$.*

Proof. To prove this we use the obvious fact that if

$$K = K_1 + K_2,$$

where K_i are non-negative bounded operators and at least one of them is invertible, then the operator K is also invertible.

Let

$$K_1 = \int\limits_0^{b_1} w_1^*(t, \lambda)w_1(t, \lambda)\,dt, \qquad K_2 = \int\limits_{b_1}^b w_1^*(t, \lambda)w_1(t, \lambda)\,dt$$

(the number b_1 will be chosen later). These operators are bounded and non-negative, and

$$\mathcal{I}_\lambda = K_1 + K_2.$$

In addition, the operator K_1 is invertible. Indeed,

$$(K_1 f, f)_\mathcal{H} = \left(\int\limits_0^{b_1} w_1^*(t, \lambda)w_1(t, \lambda)\,dt\, f, f\right)_\mathcal{H}$$

$$= \left(\int\limits_0^{b_1} \cos\sqrt{A - \overline{\lambda}I}\,t\,\cos\sqrt{A - \overline{\lambda}I}\,t\,dt\, f, f\right)_\mathcal{H}$$

$$+ \left(\int\limits_0^{b_1}\left(\int\limits_0^t w_1^*(x, \lambda)q(x)\frac{\sin\sqrt{A - \overline{\lambda}I}\,(t - x)}{\sqrt{A - \overline{\lambda}I}}\,dx\right)\cos\sqrt{A - \overline{\lambda}I}\,t\,dt\, f, f\right)_\mathcal{H}$$

$$+ \left(\int\limits_0^{b_1}\cos\sqrt{A - \overline{\lambda}I}\,t\,\left(\int\limits_0^t \frac{\sin\sqrt{A - \overline{\lambda}I}\,(t - x)}{\sqrt{A - \overline{\lambda}I}}\,q(x)w_1(x, \lambda)\,dx\right)dt\, f, f\right)_\mathcal{H}$$

$$+ \left(\int_0^{b_1} \left(\int_0^t \omega_1^*(x,\lambda) q(x) \, \frac{\sin \sqrt{A - \bar{\lambda} I}\,(t-x)}{\sqrt{A - \bar{\lambda} I}} \, dx \right) \right.$$

$$\times \int_0^t \frac{\sin \sqrt{A - \bar{\lambda} I}\,(t-x)}{\sqrt{A - \bar{\lambda} I}} \, q(x) \omega_1(x,\lambda) \, dx \right) dt \, f, f \Big)_{\mathcal{H}}$$

$$= (Tf,f)_{\mathcal{H}} + (Sf,f)_{\mathcal{H}} \,,$$

where

$$T = \int_0^{b_1} \cos \sqrt{A - \bar{\lambda} I}\, t \, \cos \sqrt{A - \lambda I}\, t \, dt \,, \quad S = K_1 - T \,.$$

It is not difficult to verify that

$$\|S\| \le c b_1^2 \,, \quad c = \text{const}\,.$$

Moreover, the constant c does not depend on b_1. Setting $\sqrt{\mu - \lambda} = \sigma + i\tau$, we obtain

$$\left(\int_0^{b_1} \cos \sqrt{A - \bar{\lambda} I}\, t \, \cos \sqrt{A - \lambda I}\, t \, dt \, f, f \right)_{\mathcal{H}}$$

$$= \int_0^{b_1} \int_0^\infty \cos \sqrt{\mu - \bar{\lambda}}\, t \, \cos \sqrt{\mu - \lambda}\, t \, d(E_\mu f, f)_{\mathcal{H}} \, dt$$

$$= \int_0^\infty \int_0^{b_1} \cos \sqrt{\mu - \bar{\lambda}}\, t \, \cos \sqrt{\mu - \lambda}\, t \, dt \, d(E_\mu f, f)_{\mathcal{H}}$$

$$= \frac{1}{2} \int_0^\infty \left(\frac{\sin(2\sigma b_1)}{2\sigma} + \frac{\sin(2i\tau b_1)}{2i\tau} \right) d(E_\mu f, f)_{\mathcal{H}}$$

$$= \frac{1}{2} \int_0^\infty \left(\frac{b_1 \sin(2\sigma b_1)}{2\sigma b_1} + \frac{b_1 \sinh(2\tau b_1)}{2\tau b_1} \right) d(E_\mu f, f)_{\mathcal{H}} \,.$$

Taking into account the estimates

$$\frac{\sin(2\sigma b_1)}{2\sigma b_1} > 0 \quad \text{if} \quad 0 < 2\sigma b_1 < \pi \,, \qquad \left| \frac{\sin(2\sigma b_1)}{2\sigma b_1} \right| < \frac{1}{\pi} \quad \text{if} \quad 2\sigma b_1 > \pi \,,$$

and

$$\frac{\sinh(2\tau b_1)}{2\tau b_1} \ge 1 \,,$$

we conclude that

$$(Tf, f)_{\mathcal{H}} \geq c_1 b_1 (f, f)_{\mathcal{H}},$$

where the positive constant c_1 does not depend on b_1. Thus, the operator T is invertible, $\|T^{-1}\| \leq cb_1^{-1}$, and $\|T^{-1}S\| \leq c_1 c^{-1} b_1 = c_2 b_1$.

Now choose b_1 so that $c_2 b_1 < 1$. Then the operator $I + T^{-1}S$ is invertible. Hence, the operator $K_1 = T + S = T(I + T^{-1}S)$ is invertible, too. □

2.4 We show here that the minimal operator L_0 generated in $\mathfrak{H} = L_2(\mathcal{H}, [0, b])$ by the operator-differential expression (2.1) and the boundary condition (2.2) is entire with respect to a certain generalized gauge in the sense of subsection A1.4. To do this we need the following result.

Lemma 2.4 *The estimate*

$$\|y(0)\|_{\mathcal{H}} \leq c\,\|y\|_+, \quad c = \text{const}, \tag{2.20}$$

holds for any vector-function $y(t) \in \mathfrak{H}_+ = \mathcal{D}(L_0^*)$.

Proof. By Lemma 2.2, a vector-function $y(t) \in \mathfrak{H}_+$ can be represented in the form

$$y(t) = \omega_1(t, 0)f + \int_0^t \omega_2(t, x, 0)\, y^*(x)\, dx, \quad f \in \mathcal{H},$$

where $y^* = L_0^* y$. It follows that

$$(\mathcal{I}_0 f, f)^{1/2} = \|\omega_1(t, 0)f\| \leq \|y\| + \left\| \int_0^t \omega_2(t, x, 0)\, y^*(x)\, dx \right\|$$

$$\leq \|y\| + c_1 \|y^*\| \leq c_2 \|y\|_+, \quad c_1, c_2 = \text{const}.$$

Since the operator $\mathcal{I}_0$ is invertible, the inequality (2.20) is true. □

It should be noted that the estimate (2.20) refers not only to the point $t = 0$ but to any other point $t \in [0, b]$. Moreover, the constant c does not depend on the choice of this point. Thus,

$$\mathfrak{H}_+ \subset C(\mathcal{H}, [0, b]),$$

where $C(\mathcal{H}, [0, b])$ denotes the space of all continuous vector-functions on the interval $[0, b]$ with values in $\mathcal{H}$. Therefore, the space $\mathfrak{H}_-$ contains the set M of the elements $\delta_0 f$, $f \in \mathcal{H}$, determined by the relation

$$(\delta_0 f, y) = (f, y(0))_{\mathcal{H}}, \quad y \in \mathfrak{H}_+.$$

Lemma 2.5 *The set M is a closed subspace in $\mathfrak{H}_-$. The operation*

$$f \mapsto \delta_0 f$$

from $\mathcal{H}$ into $\mathfrak{H}_-$ is continuous.

Proof. Let
$$y_n = \delta_0 f_n \to y \quad \text{as } n \to \infty$$
in the $\mathfrak{H}_-$-norm. Then
$$(\delta_0 f_n - \delta_0 f_m, z) = (f_n - f_m, z(0)) \to 0 \quad \text{as } n, m \to \infty$$
for any $z(t) \in \mathfrak{H}_+$. Taking $z(t)$ of the form
$$z(t) = \omega_1(t, \lambda)g, \quad g \in \mathcal{H},$$
we obtain
$$(f_n - f_m, g)_\mathcal{H} \to 0 \quad \text{as } m, n \to \infty.$$
Thus, f_n is a Cauchy sequence in the weak sense. Denote by f its weak limit. The equalities
$$(y, z) = \lim_{n \to \infty} (\delta_0 f_n, z) = \lim_{n \to \infty} (f_n, z(0))_\mathcal{H} = (f, z(0))_\mathcal{H} = (\delta_0 f, z)$$
imply that $y = \delta_0 f \in M$. So M is closed in $\mathfrak{H}_-$-topology.

By Lemma 2.4,
$$|(\delta_0 f, z)| = |(f, z(0))_\mathcal{H}| \leq \|f\|_\mathcal{H} \|z(0)\|_\mathcal{H} \leq c\|f\|_\mathcal{H} \|z\|_+,$$
hence,
$$\|\delta_0 f\|_- \leq c\|f\|_\mathcal{H}. \qquad \square$$

Lemma 2.6 *For any complex number λ the decompositions*
$$\mathfrak{H}_- = \widehat{\mathfrak{M}}_\lambda \dotplus \mathfrak{N}_{\bar\lambda}; \tag{2.21}$$
and
$$\mathfrak{H}_- = \widehat{\mathfrak{M}}_\lambda \dotplus M \tag{2.22}$$
are valid.

Proof. In view of the equality
$$\|y\|_+^2 = \|L_0^* y\|^2 + \|y\|^2 = (1 + |\lambda|^2)\|y\|^2, \quad y \in \mathfrak{N}_\lambda,$$
the norms $\|\cdot\|_+$ and $\|\cdot\|$ are equivalent on the subspace $\mathfrak{N}_\lambda$. Then the set $\mathfrak{N}_\lambda$ is closed in $\mathfrak{H}_-$, because it is closed in $\mathfrak{H}$ and in $\mathfrak{H}_+$.

To prove (2.21) suppose $y \in \widehat{\mathfrak{M}}_\lambda \cap \mathfrak{N}_{\bar\lambda}$. Since
$$y = \lim_{n \to \infty} y_n, \quad y_n \in \mathfrak{M}_\lambda,$$
(the limit is taken in the $\mathfrak{H}_-$-norm),
$$(y, z) = \lim_{n \to \infty} (y_n, z) = 0$$

for any vector-function $z(t) \in \mathfrak{N}_{\overline{\lambda}}$. Taking into account that $y \in \mathfrak{N}_{\overline{\lambda}}$, we get $(y, y) = 0$, i.e. $y = 0$.

Now let $y \in \mathfrak{H}_-$. Then

$$y = \lim_{n \to \infty} y_n, \quad y_n \in \mathfrak{H},$$

(the limit is understood in $\mathfrak{H}_-$-norm) and

$$\left(y_n - y_m, \omega_1(\cdot, \overline{\lambda})g\right) = \left(y_n{}^{\mathfrak{N}_{\overline{\lambda}}} - y_m{}^{\mathfrak{N}_{\overline{\lambda}}}, \omega_1(\cdot, \overline{\lambda})g\right) \to 0 \quad \text{as } n, m \to \infty,$$

where $y_n{}^{\mathfrak{N}_{\overline{\lambda}}} \in \mathfrak{N}_{\overline{\lambda}}$ is the component of the vector y_n in its decomposition

$$y_n = y_n{}^{\mathfrak{M}_\lambda} + y_n{}^{\mathfrak{N}_{\overline{\lambda}}}, \quad y_n{}^{\mathfrak{M}_\lambda} \in \mathfrak{M}_\lambda, \quad y_n{}^{\mathfrak{N}_{\overline{\lambda}}} \in \mathfrak{N}_{\overline{\lambda}}. \tag{2.23}$$

By virtue of $y_n{}^{\mathfrak{N}_{\overline{\lambda}}} = \omega_1(t, \overline{\lambda})f_n$, $f_n \in \mathcal{H}$,

$$\left(\omega_1(t, \overline{\lambda})(f_n - f_m), \omega_1(t, \overline{\lambda})g\right) = \left(\mathcal{I}_{\overline{\lambda}}(f_n - f_m), g\right)_{\mathcal{H}} \to 0 \quad \text{as } n, m \to \infty.$$

Since $\mathcal{I}_{\overline{\lambda}}$ is invertible, the sequence f_n converges weakly to a certain element $f \in \mathcal{H}$. Let us show that $f_n \to f$ strongly. Indeed,

$$\|\omega_1(\cdot, \overline{\lambda})f_n\|_+^2 = (1 + |\lambda|^2) \int_0^b \left(\omega_1(t, \overline{\lambda})f_n, \omega_1(t, \overline{\lambda})f_n\right)_{\mathcal{H}} dt$$

$$= (1 + |\lambda|^2)\left(\mathcal{I}_{\overline{\lambda}}f_n, f_n\right)_{\mathcal{H}} \le \alpha\|\mathcal{I}_{\overline{\lambda}}\|, \quad \alpha = \text{const}.$$

So,

$$\left|\left(y_n - y_m, \omega_1(\cdot, \overline{\lambda})(f_n - f_m)\right)\right| \le \|y_n - y_m\|_- \cdot \|\omega_1(\cdot, \overline{\lambda})(f_n - f_m)\|_+$$
$$\le \alpha\|y_n - y_m\|_- \cdot \|\mathcal{I}_{\overline{\lambda}}\| \to 0 \quad \text{as } n, m \to \infty.$$

Therefore

$$\left(\mathcal{I}_{\overline{\lambda}}(f_n - f_m), f_n - f_m\right)_{\mathcal{H}} = \left(y_n - y_m, \omega_1(\cdot, \overline{\lambda})(f_n - f_m)\right) \to 0 \quad \text{as } n, m \to \infty.$$

But the operator $\mathcal{I}_{\overline{\lambda}}$ is invertible, hence $\|f_n - f_m\| \to 0$ when $n, m \to \infty$, i.e. $f_n \to f$ strongly.

It follows from the equality

$$\left\|y_n{}^{\mathfrak{N}_{\overline{\lambda}}} - y_m{}^{\mathfrak{N}_{\overline{\lambda}}}\right\|^2 = \left(\mathcal{I}_{\overline{\lambda}}(f_n - f_m), f_n - f_m\right)_{\mathcal{H}}$$

that $y_n{}^{\mathfrak{N}_{\overline{\lambda}}}$ tends to a certain element $y^{\mathfrak{N}_{\overline{\lambda}}}$. In view of the closedness of the set $\mathfrak{N}_{\overline{\lambda}}$ in $\mathfrak{H}_-$, $y^{\mathfrak{N}_{\overline{\lambda}}} \in \mathfrak{N}_{\overline{\lambda}}$. The relation $\overline{\mathfrak{M}_\lambda} = \widehat{\mathfrak{M}}_\lambda$ allows us to conclude that $y_n{}^{\mathfrak{M}_\lambda} \to y^{\mathfrak{M}_\lambda} \in \widehat{\mathfrak{M}}_\lambda$. Passing to the limit in (2.23) as $n \to \infty$, we get the decomposition (2.21).

Assume now $y \in M \cap \widehat{\mathfrak{M}}_\lambda$. Thus, $y = \delta_0 f \in \widehat{\mathfrak{M}}_\lambda$. Since $\overline{\mathfrak{M}_\lambda} = \widehat{\mathfrak{M}}_\lambda$ in $\mathfrak{H}_-$, we obtain, by using (2.21),

$$(y, \omega_1(\cdot, \overline{\lambda})g) = (\delta_0 f, \omega_1(\cdot, \overline{\lambda})g) = (f, g)_{\mathcal{H}} = 0$$

for any $g \in \mathcal{H}$. So $f = 0$, whence $y = 0$. Consequently,

$$M \cap \widehat{\mathfrak{M}}_\lambda = 0.$$

Let y be an arbitrary vector from $\mathfrak{H}_-$. Choose the element $f \in \mathcal{H}$ so that

$$(f, g)_{\mathcal{H}} = (y, \omega_1(\cdot, \overline{\lambda})g)$$

for any $g \in \mathcal{H}$ (by the Riesz theorem on the general form of a continuous linear functional the vector f is uniquely determined). Then

$$(y - \delta_0 f, \omega_1(\cdot, \overline{\lambda})g) = (y, \omega_1(\cdot, \overline{\lambda})g) - (f, g)_{\mathcal{H}} = 0.$$

The decomposition (2.21) shows that $z = y - \delta_0 f \in \widehat{\mathfrak{M}}_\lambda$. Hence, $y = \delta_0 f + z$ which proves the validity of the decomposition (2.22). $\qquad\square$

Assign to each $y \in \mathfrak{H}$ the vector-function

$$\tilde{y}(\lambda) = \int_0^b \omega_1^*(t, \overline{\lambda})\, y(t)\, dt. \tag{2.24}$$

The transform

$$y \mapsto \tilde{y}(\lambda) \tag{2.25}$$

possesses the properties:

a) for a fixed λ it maps $\mathfrak{H}$ into $\mathcal{H}$ continuously;

b) for a fixed $y \in \mathfrak{H}$, $\tilde{y}(\lambda)$ is an entire vector-function with values in $\mathcal{H}$;

c) if $y \in \mathfrak{M}_{\lambda_0}$, then $\tilde{y}(\lambda_0) = 0$.

Lemma 2.7 *The transform (2.25) can be extended to the whole space $\mathfrak{H}_-$ so that the properties* a), b) *and* c) *are preserved.*

Proof. For any $y \in \mathfrak{H}$ we have

$$(\tilde{y}(\lambda), f)_{\mathcal{H}} = \int_0^b \left(\omega_1^*(t, \overline{\lambda})\, y(t), f\right)_{\mathcal{H}} dt = (y, \omega_1(\cdot, \overline{\lambda})f), \quad f \in \mathcal{H}. \tag{2.26}$$

However, the expression $(y, \omega_1(\cdot, \overline{\lambda})f)$ is meaningful for vector-functions $y(t) \in \mathfrak{H}_-$. Moreover,

$$\left|(y, \omega_1(\cdot, \overline{\lambda})f)\right| \leq \|y\|_- \cdot \|\omega_1(\cdot, \overline{\lambda})f\|_+$$

$$= \sqrt{1 + |\lambda|^2} \, \|y\|_- \cdot \|\omega_1(\cdot, \overline{\lambda})f\| \tag{2.27}$$

$$\leq \sqrt{1 + |\lambda|^2} \, \|y\|_- \cdot \|\mathcal{I}_\lambda\|^{1/2} \cdot \|f\|_{\mathcal{H}} \, .$$

This inequality shows that for fixed λ and $y \in \mathfrak{H}_-$ the scalar product $(y, \omega_1(\cdot, \overline{\lambda})f)$ gives an antilinear continuous functional on the space $\mathcal{H}$. By the Riesz theorem,

$$(y, \omega_1(\cdot, \overline{\lambda})f) = (\widetilde{y}(\lambda), f)_{\mathcal{H}} \, .$$

As is seen from (2.26), $\widetilde{y}(\lambda)$ coincides with that in the formula (2.24) if $y \in \mathfrak{H}$. Thus, the transform (2.24) is continuously extended from $\mathfrak{H}$ to $\mathfrak{H}_-$. Since

$$\mathfrak{H}_- \ni y = \lim_{n \to \infty} y_n \, , \quad y_n \in \mathfrak{H} \, ,$$

(the limit is taken in the $\mathfrak{H}_-$-norm), the sequence of entire vector-functions $\widetilde{y}_n(\lambda)$ converges, due to the estimate (2.27), uniformly in any bounded domain of the complex plane. So, the function $\widetilde{y}(\lambda)$ is also entire. Finally, the property c) follows from the relation $\overline{\mathfrak{M}_\lambda} = \widehat{\mathfrak{M}_\lambda}$ in $\mathfrak{H}_-$-topology. $\square$

Theorem 2.3 *The operator L_0 is entire with respect to the generalized gauge $M = \{\delta_0 f\}_{f \in \mathcal{H}}$. The operator-function $\mathcal{P}(\lambda)$ of the first kind is of the form*

$$\mathcal{P}(\lambda)\, y = \delta_0\, \widetilde{y}(\lambda) \, , \quad y \in \mathfrak{H}_- \, . \tag{2.28}$$

Proof. The fact that the operator L_0 is M-entire follows from Corollary 2.5 and the decomposition (2.22). Multiplying scalarly by $\omega_1(t, \overline{\lambda})g \in \mathfrak{N}_{\overline{\lambda}} \subset \mathfrak{H}_+, g \in \mathcal{H}$, the equality

$$y = y^{\widehat{\mathfrak{M}_\lambda}} + \delta_0 f \, , \quad \text{we get} \quad (\widetilde{y}(\lambda), g)_{\mathcal{H}} = (f, g)_{\mathcal{H}} \, , \quad \text{whence} \quad f = \widetilde{y}(\lambda)$$

which completes the proof. $\square$

Using the representation (2.19) for $y = (L_0 - \lambda I)^{-1}h$ and the formula (2.28), one can show that the operator-function $T(\lambda)$ of the second kind admits the representation

$$T(\lambda)y = \int_t^b \left(\omega_1(t, \lambda)\, \omega_2^*(\xi, \overline{\lambda}) - \omega_2(t, \lambda)\, \omega_1^*(\xi, \overline{\lambda})\right) y(\xi)\, d\xi \, .$$

2.5   A certain operator generalization of the directing functional notion introduced in section 2.8 is given in this subsection.

Let $\mathfrak{L}$ be a linear set with a quasiscalar product $(\cdot, \cdot)_{\mathfrak{L}}$.

Definition 2.2 *The map*

$$\Phi : (\lambda, u) \mapsto \Phi_\lambda u$$

from $\mathbb{R}^1 \times \mathfrak{L}$ into a Hilbert space $\mathcal{H}$ is called a directing "functional" for a Hermitian densely defined operator T on $\mathfrak{L}$ if:

1) *Φ is linear in u;*

2) *for every fixed $u \in \mathfrak{L}$ the vector-function $\Phi_\lambda u$ is analytic on $\mathbb{R}^1$ in λ;*

3) *for any $\lambda \in \mathbb{R}^1$, any $u \in \mathfrak{L}$ the equation*

$$(T - \lambda I)\, v = u$$

has a solution $v \in \mathcal{D}(T)$ if and only if $\Phi_\lambda u = 0$;

4) *for an arbitrary closed interval $\Delta \subset \mathbb{R}^1$ there exists a map*

$$\Psi^{(\Delta)} : (\lambda, f) \mapsto \Psi_\lambda^{(\Delta)} f$$

from $\Delta \times \mathcal{H}$ into $\mathfrak{L}$ which possesses the properties:

a) *$\Psi^{(\Delta)}$ is linear in f for any fixed $\lambda \in \Delta$;*

b) *for each $f \in \mathcal{H}$ the vector-function $\Psi_\lambda^{(\Delta)} f$ is analytic on $\mathbb{R}^1$;*

c) *$\Phi_\lambda\big(\Psi_\lambda^{(\Delta)} f\big) = f$;*

d) *the operator $\widehat{\Psi}^{(\Delta)}$ from $\mathcal{H}$ into $\widehat{\mathfrak{L}}$ is continuous (recall that $\widehat{\mathfrak{L}}$ denotes the space obtained by identification with zero of all elements from $\mathfrak{L}$ for which $(u, u)_\mathfrak{L} = 0$, and $\widehat{\Psi}_\lambda^{(\Delta)} f$ is the class corresponding to $\Psi_\lambda^{(\Delta)} f$ under this identification).*

By analyticity in λ of the function $\Psi_\lambda^{(\Delta)} f$ (f is fixed) we mean that of the function $\widehat{\Psi}_\lambda^{(\Delta)} f$ in the Hilbert space $\widehat{\mathfrak{L}}$.

The next theorem was established by Langer [1].

Theorem 2.4 *Let a Hermitian operator T on $\mathfrak{L}$ have a directing "functional" Φ with values in a Hilbert space $\mathcal{H}$. For every spectral function E_λ of the operator T there exists a non-decreasing left-continuous operator-function $\rho(\lambda) : \rho(0) = 0$ on $\mathbb{R}^1$ with values in $L[\mathcal{H}]$ such that*

$$\big(E_\Delta u, v\big)_\mathfrak{L} = \int\limits_\Delta \big(d\rho(t)\Phi_t u, \Phi_t v\big)_\mathcal{H}$$

for any $u, v \in \mathfrak{L}$.

We choose the space $\mathfrak{H} = L_2(\mathcal{H}, [0, b])$, $b < \infty$, as $\mathfrak{L}$ and the minimal operator L_0 generated on $\mathfrak{H}$ by the operator differential expression (2.1) and the boundary condition (2.2) as T. The following assertion is valid.

Lemma 2.8 *The "functional"*

$$\Phi_\lambda y = \tilde{y}(\lambda) = \int_0^b \omega_1^*(t, \overline{\lambda}) \, y(t) \, dt \, , \quad y \in \mathfrak{H} \tag{2.29}$$

is directing for the minimal operator L_0 in the sense of Definition 2.2.

Proof. Properties 1) and 2) of a directing "functional" are obvious. To prove property 3) suppose $z \in \mathcal{D}(L_0)$ to be a solution of the equation

$$(L_0 - \lambda I) \, z = y \, , \quad \lambda \in \mathbb{R}^1, \quad y \in \mathfrak{H} \, . \tag{2.30}$$

Multiply scalarly by $\omega_1(t, \overline{\lambda}) f$, $f \in \mathcal{D}(\mathcal{A})$, the identity (2.30). We have

$$\int_0^b \big((L_0 - \lambda I) \, z, \omega_1(t, \overline{\lambda}) f \big)_{\mathcal{H}} \, dt = \int_0^b \big(y(t), \omega_1(t, \overline{\lambda}) f \big)_{\mathcal{H}} \, dt \, .$$

Taking into account that $\omega_1(t, \overline{\lambda}) f$ is a solution of the equation $l[y] = \overline{\lambda} y$, we find, by using the integration-by-parts formula (2.6), that

$$\int_0^b \big(\omega_1^*(t, \overline{\lambda}) \, y(t), f \big)_{\mathcal{H}} \, dt = 0$$

for any $f \in \mathcal{D}(\mathcal{A})$, whence $\Phi_\lambda y = 0$.
   Conversely, let

$$\Phi_\lambda y = \int_0^b \omega_1^*(t, \overline{\lambda}) \, y(t) \, dt = 0 \tag{2.31}$$

for some $y \in \mathfrak{H}, \lambda \in \mathbb{R}^1$. It is not difficult to verify that the vector-function

$$z(t) = \int_t^b \omega_2(t, x, \lambda) \, y(x) \, dx$$

satisfies the equation

$$\hat{l}[z] - \lambda z = y \tag{2.32}$$

and the boundary conditions

$$z(b) = z'(b) = 0 \, .$$

Multiplying the equality (2.31) scalarly by $f \in \mathcal{D}(\mathcal{A})$, we obtain, by integrating by parts with regard to the formulas (2.6) and (2.32), that

$$0 = \int_0^b (\omega_1(x, \overline{\lambda})f, y(x))_{\mathcal{H}} \, dx = (\omega_1(0, \overline{\lambda})f, z'(0))_{\mathcal{H}} = (f, z'(0))_{\mathcal{H}},$$

whence $z'(0) = 0$. Thus, $z \in \mathcal{D}(L_0)$ and $y = (L_0 - \lambda I)z$.

Now we shall show that the "functional" (2.29) possesses the property 4). Let $\psi(t)$ be an infinitely differentiable non-negative scalar function such that for some $a_1, a_2 : 0 < a_1 < a_2 < b$

$$\psi(t) = \begin{cases} 1 & \text{as} \quad 0 < t < a_1 \\ 0 & \text{as} \quad a_2 < t \leq b. \end{cases}$$

Set

$$\mathcal{I}_{\lambda\psi} = \int_0^b \psi(t) \, \omega_1^*(t, \overline{\lambda}) \, \omega_1(t, \lambda) \, dt.$$

The operator-function $\mathcal{I}_{\lambda\psi}$ is entire. Since for $\lambda \in \mathbb{R}^1$ the operator $\mathcal{I}_{\lambda\psi}$ can be represented as the sum

$$\int_0^{a_1} \omega_1^*(t, \overline{\lambda}) \, \omega_1(t, \lambda) \, dt + \int_{a_1}^b \psi(t) \, \omega_1^*(t, \overline{\lambda}) \, \omega_1(t, \lambda) \, dt$$

of two non-negative operators, in which the first summand is, by Lemma 2.3, invertible, the operator $\mathcal{I}_{\lambda\psi}$ is invertible, too. In addition, the operator-function $\mathcal{I}_{\lambda\psi}^{-1}$ is analytic on $\mathbb{R}^1$.

We define the operators $\Psi_\lambda^{(\Delta)}$ appearing in the property 4) as follows:

$$\Psi_\lambda^{(\Delta)}f = \Psi_\lambda f = \psi(t)\,\omega_1(t, \lambda)\,\mathcal{I}_{\lambda\psi}^{-1}f, \quad f \in \mathcal{H}.$$

The operator $\Psi_\lambda^{(\Delta)}$ acts continuously from $\mathbb{R}^1 \times \mathcal{H}$ into $\mathfrak{H}$. Moreover, for a fixed f the vector-function $\Psi_\lambda^{(\Delta)}f$ is analytic on $\mathbb{R}^1$. Further,

$$\Phi_\lambda\Psi_\lambda^{(\Delta)}f = \int_0^b \omega_1^*(t, \lambda)\,\psi(t)\omega_1(t, \lambda)\,\mathcal{I}_{\lambda\psi}^{-1}f \, dt = \mathcal{I}_{\lambda\psi}\,\mathcal{I}_{\lambda\psi}^{-1}f = f.$$

Thus, the operator $\Psi_\lambda^{(\Delta)}$ possesses all the properties a)–d) in 4). Consequently, Φ is a directing "functional" for the Hermitian operator L_0. $\qquad\square$

Lemma 2.8 and Theorem 2.4 lead to the following result.

Theorem 2.5 *For every spectral function E_λ of the minimal operator L_0 there exists a non-decreasing left-continuous operator-function $\rho(\lambda)$, $\lambda \in \mathbb{R}^1$, normalized by the condition $\rho(0) = 0$, such that the equality*

$$\int_0^b \left(E_\Delta \, y(t), z(t) \right)_{\mathcal{H}} dt = \int_\Delta \left(d\rho(\lambda) \, \tilde{y}(\lambda), \tilde{z}(\lambda) \right)_{\mathcal{H}}, \qquad (2.33)$$

is valid for an arbitrary interval $\Delta \subseteq \mathbb{R}^1$ and any $y, z \in \mathfrak{H} = L_2(\mathcal{H}, [0, b])$.

If we choose $\Delta = \mathbb{R}^1$, we arrive at the Parseval equality

$$(y, z) = \int_{-\infty}^{\infty} \left(d\rho(\lambda) \, \tilde{y}(\lambda), \tilde{z}(\lambda) \right)_{\mathcal{H}}.$$

The function $\rho(\lambda)$ is called a spectral function of the problem (2.4),(2.2). Since the defect numbers of the operator L_0 are infinite, the set of the functions $\rho(\lambda)$ giving the Parseval equality is infinite. Taking a finite interval Δ in (2.33), the functions

$$\chi_{nf}(t) = \begin{cases} nf & \text{if} \quad t \in [0, \frac{1}{n}] \\ 0 & \text{if} \quad t \in (\frac{1}{n}, b] \end{cases}, \qquad f \in \mathcal{H},$$

as $y(t)$ and $\chi_{ng}(t)$, $g \in \mathcal{H}$, as $z(t)$, we obtain, by passing to the limit when $n \to \infty$, that

$$\left(\widehat{E}_\Delta \delta_0 \, f, \delta_0 \, g \right) = \left(\rho(\Delta) f, g \right)_{\mathcal{H}}.$$

The relation (A1.23) allows us to conclude that there exists a one-to-one correspondence between the set of all the functions $\rho(\lambda)$ and the set of all the functions $\sigma(\lambda)$ whose complete description is given by formula (A1.26). Going from the latter, it is not hard to describe the set of all spectral functions of the problem (2.4),(2.2). This description is given by the next theorem.

Theorem 2.6 *The formula*

$$\left(W_{22}(z)\tau(z) + W_{21}(z) \right) \left(W_{12}(z)\tau(z) + W_{11}(z) \right)^{-1}$$

$$= C + \int_{-\infty}^{\infty} \left(\frac{1}{\lambda - z} - \frac{\lambda}{\lambda^2 + 1} \right) d\rho(\lambda), \qquad \Im z > 0,$$

$(C \in L(\mathcal{H}))$ determines a one-to-one correspondence between the set of all spectral functions of the problem (2.4),(2.2) and the set of all functions

$$\tau(z) = i \left(I_M + \Omega(z) \right) \left(I_M - \Omega(z) \right)^{-1},$$

where $\Omega(z)$ runs through the set of all analytic operator-functions in the half-plane $\Im z > 0$ whose values are contractions on the space $\mathcal{H}$ $(\|\Omega(z)\| \leq 1)$. Here

$$W_{11}(z) = 1 + z \int_0^b w_1^*(t, \bar{z}) \, w_2(t, 0) \, dt \, ,$$

$$W_{12}(z) = -z \int_0^b w_1^*(t, \bar{z}) \, w_1(t, 0) \, dt \, ,$$

$$W_{21}(z) = z \int_0^b w_2^*(t, \bar{z}) \, w_2(t, 0) \, dt \, ,$$

$$W_{22}(z) = 1 - z \int_0^b w_2^*(t, \bar{z}) \, w_1(t, 0) \, dt \, .$$

Appendix 3

On a Difficult Problem of Operator Theory and its Relation to Classical Analysis

(M.G. Krein's talk at the Jubilee scientific session
of the Moscow Mathematical Society,
Moscow 1964)

I must admit that the title of my talk is somewhat indeterminate and it should be specified. The issue to discuss here is certain representations (I propose to call them canonical) of self-adjoint operators. In mathematical physics such operators usually appear in the form of the self-adjoint differential expressions with some self-adjoint boundary conditions added. The aim of the theory I shall talk about is to clarify whether this is accidental. More exactly, is it derived only from the nature of the problems suggested by technology and natural sciences or is it based on the purely logical necessity to represent any self-adjoint operator as a differential one? If the latter is true, then where does this necessity arise from? We shall consider the simplest of infinite-dimensional Banach spaces, a separable Hilbert space, and the simplest, with respect to spectral properties, unbounded operators on it, self-adjoint ones. The most elementary of them are the operators whose spectrum is simple.

By the 1930s, after the works by T. Carleman, J. von Neumann, and M. Stone, the foundations of the theory of Hermitian, in particular, self-adjoint operators had been constructed. It remained to add quite a little, using the results by E. Hellinger, H. Hahn, and the integration theory (T. Stieltjes, H. Lebesgue, J. Radon) to obtain the spectral types theory for self-adjoint operators. The latter is perfectly presented in the monograph by A. Plesner and V. Rokhlin. Complete clarity in the theory of generalized spectral functions came with M. Naimark's works. The theory of Hermitian operators seemed to be completed, and there was a feeling that for future researchers it remained only to apply this completed theory to related fields of mathematics and physics. But in reality, difficult problems requiring new analytical techniques were appearing on the horizon. The problem I am going to tell about is one of those observed lately. Several decades passed before it was formulated in the case of a self-adjoint operator whose spectrum is simple. Only a couple years ago was this problem finally solved. However there is no complete solution in the case of multiple spectrum. As for operators whose spectrum is infinite multiplicity, we are, so to say, wandering in darkness. What is the problem indeed?

To give the idea, consider a Hermitian operator A with a simple spectrum on a finite-dimensional space $\mathfrak{H}$:

$$\dim \mathfrak{H} = n < \infty.$$

The simplicity of the spectrum in this case means all the roots of the characteristic equation of the symmetric matrix representing A being different. But everyone involved in numerical computations knows that in order to check the simplicity of the spectrum of A, it suffices to verify that the vectors

$$u, \, Au, \, \ldots, \, A^{n-1}u$$

form a basis in $\mathfrak{H}$ for some $u \in \mathfrak{H}$. Such a vector u is called a generating one. It should be noted that the number of the generating vectors is infinite for each Hermitian operator on $\mathfrak{H}$ with a simple spectrum. Any vector $u \in \mathfrak{H}$ with nonzero projections to all eigenvectors of this operator is a generating one.

Having fixed a generating vector u, we put

$$\mathfrak{H}_k = \mathrm{span}\big\{u, Au, \ldots, A^{k-1}u\big\}$$

and define the operator $\widehat{A}_k$ in the following way:

$$\widehat{A}_k f = A f, \qquad \mathcal{D}(\widehat{A}_k) = \mathfrak{H}_{k-1}$$

($\mathcal{D}(\cdot)$ denotes the domain of an operator). Considering $\widehat{A}_k$ as an operator on the space $\mathfrak{H}_k$, we can see that the dimension of its domain as well as that of its range differs from $\dim \mathfrak{H}_k$ by 1. Thus, we have the chain of the spaces

$$\mathfrak{H}_1 \subset \mathfrak{H}_2 \subset \cdots \subset \mathfrak{H}_{k-1} \subset \mathfrak{H}_k \subset \cdots \subset \mathfrak{H}_n = \mathfrak{H}$$

and the set of the Hermitian operators $\widehat{A}_k$ such that the deficiencies of any two neighbouring operators differ by 1.

Choose a vector $\varphi_k \in \mathfrak{H}_k$ such that

$$\varphi_k \perp \mathfrak{H}_{k-1}.$$

In other words, we carry out the Schmidt orthogonalization procedure for the sequence $u, Au, \ldots, A^{n-1}u$. The coefficients in this procedure may be taken as real. Then the element $A\varphi_k$ can be written as

$$A\varphi_k = -b_{k-1}\varphi_{k-1} + a_k\varphi_k - b_k\varphi_{k+1}, \quad k = 1, 2, \ldots, n-1,$$

where

$$\Im a_k = 0, \, b_0 = b_{n+1} = 0, \, b_k > 0, \, k = 1, 2, \ldots, n-1.$$

So, in the basis $\{\varphi_k\}_{k=1}^{n}$ the operator A is represented by a Hermitian Jacobi matrix.

Summarizing what has been said concerning the finite-dimensional case, I would like to draw attention to the following fact: if we took the eigenvectors of A as a basis, we would represent the operator A by the diagonal matrix. But if we fix only one axis and choose the others so that the matrix of the operator A is of the simplest form, we can obtain the representation of A by a Jacobi matrix. If, in addition, the operator A is positive, the characterization of this representation can be given in more detail. But just now I am not concentrating on this because below I shall describe a more general situation.

Let us consider the case where

$$\dim \mathfrak{H} = \infty.$$

Suppose first a Hermitian operator A on $\mathfrak{H}$ to be bounded. As in the finite-dimensional case, one can define an operator with a simple spectrum. Namely, the spectrum of A is said to be simple if there exists an element $u \in \mathfrak{H}$ such that

$$\overline{\mathrm{span}\{u, Au, \ldots, A^n u, \ldots\}} = \mathfrak{H}.$$

Orthogonalizing the sequence $\{A^k u\}_{k=0}^{\infty}$, we get the basis with respect to which the operator A is represented by an infinite Jacobi matrix. A lot of mathematicians have been studying such matrices for many years. In this connection the works by P. Chebyshev, A. Markov, and T. Stieltjes should be mentioned. The main tools of investigation in these works are continued fractions.

Pass now to the case of unbounded self-adjoint operators (including, for instance, the differentiation operator). In this case the situation is more complicated. Firstly, an unbounded operator A can not be applied to an arbitrary element $u \in \mathfrak{H}$. Secondly, even if all the powers $A^k u$ exist, their closed linear span may not coincide with the whole space $\mathfrak{H}$. In the situation under consideration we must give another, more refined definition of a generating vector.

As we have seen, in the finite-dimensional case a generating vector is the one having a nonzero projection to every eigenvector of A. We shall proceed from this property. We say that the spectrum of a self-adjoint operator A on $\mathfrak{H}$ is simple if

$$\overline{\mathrm{span}\{E_\lambda u\}_{\lambda \in \mathbb{R}^1}} = \mathfrak{H}$$

for some element $u \in \mathfrak{H}$; here E_λ, $\lambda \in \mathbb{R}^1$, is the spectral function of the operator A:

$$A = \int_{-\infty}^{\infty} \lambda \, dE_\lambda.$$

Such an element u is called a generating one. Set

$$\sigma(\lambda) = (E_\lambda u, u).$$

Then there exists a unitary mapping

$$f \mapsto f_u(\lambda) \tag{3.1}$$

from $\mathfrak{H}$ onto the space $L_2(\mathbb{R}^1, d\sigma)$ such that

$$Af \mapsto \lambda f_u(\lambda). \tag{3.2}$$

So as physicists would say, we obtain the diagonal representation of the operator A. But we shall be interested not in the representation (3.2) but in another one, which is the analogue of the representation in the Jacobi matrix form.

To give some perspective let us formulate the result concerning the case of a positive A, that is $\sigma(\lambda) = 0$ as $\lambda < 0$. Then a certain string "tightened by the unit force" can be associated with the operator A. This string has a ring at its left end (i.e. at the point $(0,0)$) whose mass is $M(+0) = \dfrac{1}{(u,u)} = \dfrac{1}{\sigma(\infty)}$, which slides freely along the y-axis. The right end is fixed. Here $M(x)$ denotes the mass of the part of the string from 0 to x, $x \in [0, l]$, $l \le \infty$. It turns out that the unitary map

$$f \mapsto f_M(x) \tag{3.3}$$

from $\mathfrak{H}$ onto $L_2([0, l], dM)$ exists, such that under this map the operator A is transformed into the string operator:

$$Af \mapsto -\frac{d}{dM} \cdot \frac{df_M(x)}{dx}. \tag{3.4}$$

The relation of $f_M(x)$ from (3.3) to the function $f_u(\lambda)$ from (3.1) is given by the formula

$$f_u(\lambda) = \int\limits_0^l f_M(x)\varphi(x, \lambda)\, dM(x),$$

where $\varphi(x, \lambda)$ is the solution of the equation

$$L\varphi = \lambda\varphi, \quad L = -\frac{d}{dM} \cdot \frac{d}{dx},$$

satisfying the conditions

$$\varphi(0, \lambda) = 1, \qquad \varphi'(0, \lambda) = 0.$$

Of course, one should clarify the sense of the differentiation with respect to mass if, say, the latter has constancy intervals or jump points. But all this can be done thoroughly. Besides, for the image of the operator A to be self-adjoint, certain boundary conditions must be set. In this specific case they are

$$\frac{df_M(x)}{dx}\bigg|_{x=0} = 0 \tag{3.5}$$

(free sliding of the ring along the y-axis) and

$$f_M(l) = 0 \quad \text{if} \quad l < \infty \tag{3.6}$$

(the right end being fixed). The condition (3.6) is omitted if $l = \infty$ or $M(l) = \infty$.

It is not hard to observe that the expression (3.4) is the continual analogue of the second difference, and the operator (3.4)–(3.6) may be considered as a limit of its restrictions $\widehat{A}_x$ to the "truncated" subspaces of the functions $f_M(x)$ from $L_2\big([0,l],dM\big)$ vanishing outside $[0,x]$ and satisfying the conditions (3.5) and

$$f_M(x) = 0, \quad \left.\frac{df_M(s)}{ds}\right|_{s=x} = 0.$$

These restrictions are complete analogues of the operators $\widehat{A}_k$ appearing in the finite-dimensional situation. Note further that, as in the finite-dimensional case, the number of the strings which can be associated with the given operator is infinite. Each generating element generates the corresponding string.

The above result was published in 1952 (see Dokl. Akad. Nauk SSSR, 87 (1952), No.6, 881-885). Two years later, A. Kolmogorov presented my next paper in which I showed that as soon as the operators $\widehat{A}_x$ have been constructed, the interpolation and filtration problems for stationary stochastic processes were effectively solved. I shall not concentrate on all paradoxical conclusions obtained in them if, for instance, the derivative of the function $M(x)$ equals 0 almost everywhere. Nor am I focusing your attention on the case where a generating element is generalized. We have quite precise results in this case, too. But I was very surprised when the differential operation $-\dfrac{dy'}{dM}$ considered by W. Feller in connection with the description of one-dimensional Markov birth and death processes (see Trans. Amer. Math. Soc., 77 (1954), 1–31), was called by A. Yaglom the Feller operator. The point is that this operator is a special case of the above string operator with the function $M(x)$ having no constancy intervals. If W. Feller were acquainted with my works, he would know that even the interpolation and extrapolation problems are completely solved by using the methods developed there.

Turn now to the general situation. As is known, any second order differential equation may be considered as the continual analogue of an operator given by a Jacobi matrix. The question arises whether the reserve of these differential equations is rich enough to get all the representations for all self-adjoint operators on $\mathfrak{H}$. It turns out that this reserve is not enough even if we allow the coefficients of a differential expression to be generalized functions. The class required will be found if we introduce the canonical differential operators. Using them, we can describe arbitrary self-adjoint operators, including the operators whose spectrum is multiple. The main role in this description belongs to an entire Hermitian operator. Let us give its definition.

Let $\widehat{A}$ be a simple closed densely defined Hermitian operator on $\mathfrak{H}$:

$$\big(\widehat{A}f,g\big) = \big(f,\widehat{A}g\big), \quad f,g \in \mathcal{D}(\widehat{A}).$$

Put

$$\mathfrak{M}_z = \big(\widehat{A} - zI\big)\mathcal{D}(\widehat{A}) \qquad \text{and} \qquad \mathfrak{N}_z = \mathfrak{M}_{\bar{z}}^{\perp}$$

($\perp$ is the sign of the orthogonal complement). The subspace $\mathfrak{M}_z$ is closed for every $z : \Im z \neq 0$. It is well-known that $\dim \mathfrak{N}_z$ is the same for all $z : \Im z > 0$ and for all $z : \Im z < 0$. We suppose these dimensions to be equal.

Fix a subspace $M \subset \mathfrak{H}$ (the module of the representation). We call the operator $\widehat{A}$ entire (or M-entire) if it satisfies the following two conditions:

1) the subspace $\mathfrak{M}_z$ is closed for any number $z \in \mathbb{C}^1$, in particular, for the real ones;

2) the angle between the subspaces M and $\mathfrak{N}_z$ is less than $\frac{\pi}{2}$ for any $z \in \mathbb{C}^1$.

The property 2) shows that the subspaces M and $\mathfrak{N}_z$ are projected one-to-one and mutually continuously onto each other in parallel to $\mathfrak{M}_z$.

Denote by $f_M(z)$ the oblique projection parallel to $\mathfrak{M}_z$ of a vector $f \in \mathfrak{H}$ to M. This means that

$$f - f_M(z) \in \mathfrak{M}_z .$$

If the operator $\widehat{A}$ is M-entire, the function $f_M(z)$ proves to be entire for every $f \in \mathfrak{H}$. Its growth order may be arbitrary in the case of an infinite deficiency index

$$n = \dim \mathfrak{N}_z$$

of the operator $\widehat{A}$. But if n is finite, then the growth of $f_M(z)$ is at most exponential. More exactly,

$$\varlimsup_{|z| \to \infty} \frac{\log \|f_M(z)\|}{|z|} \leq h < \infty , \qquad (3.7)$$

where the number h does not depend on the choice of the element f. The minimal value of h in (3.7) is called the type of $\widehat{A}$. The functions $f_M(z)$ possess some additional properties. In particular, they belong to the Cartwright class. To enable you to imagine a completed result, I restrict myself to the case of a finite-dimensional M, especially as in the last one it is possible to say much more.

Let $u_1, u_2, \ldots, u_n$ be a basis in M. Then the representing function $f_M(z)$ is written as

$$f_M(z) = \{f_1(z)u_1, f_2(z)u_2, \ldots, f_n(z)u_n\} .$$

The function

$$f_u(z) = \{f_1(z), f_2(z), \ldots, f_n(z)\}$$

is a n-tuple consisting of the entire scalar functions $f_k(z)$. If $n = 1$, then $f_u(z)$ is an ordinary scalar function. Under the map

$$f \mapsto f_u(z) \qquad (3.8)$$

the operator $\widehat{A}$ is transformed into the multiplication by z:

$$\widehat{A}f \mapsto z f_u(z) .$$

The map (3.8) is isometric. This means that there exists a non-decreasing $(n \times n)$-matrix-function $\sigma(\lambda)$ such that, if the scalar product in the space of entire vector-functions $f_u(z)$ is introduced as

$$\left(f_u, g_u\right)_\sigma = \int_{-\infty}^{\infty} f_u(\lambda)\, d\sigma(\lambda)\, g_u^*(\lambda)\,, \qquad (3.9)$$

where $g_u^*(\lambda)$ is the corresponding vector-column, then

$$(f, g) = \left(f_u, g_u\right)_\sigma. \qquad (3.10)$$

In the case of $n = 1$ (we call it scalar) $\sigma(\lambda)$ is an ordinary non-decreasing function of bounded variation on the whole real axis.

The number of the matrices $\sigma(\lambda)$ guaranteeing the representation (3.10) is infinite. We can give the complete description of such matrices. One can find among them, for example, the ones whose increase points are isolated (jump functions) as well as the functions of the form $\tau^{-2}(\lambda)$ where $\tau(\lambda)$ is entire, etc. It is possible, I repeat, to give the description of all such functions. The Chebyshev-Markov problem may be stated for them. Why did I recall this problem? Because finding these functions is a certain moment problem: knowing the numbers (f, g) and the functions $f_u(z), g_u(z)$, we search for a distribution function $\sigma(\lambda)$ such that the integral (3.9) take the given values. Even in the scalar case this generalized problem includes the indefinite case (the most difficult one) of the moment problem, the Stieltjes problem, the continuation problem for Hermitian-positive functions, and, among other things, the extrapolation problem for stochastic processes.

The Chebyshev-Markov problem lies in determining the maximum of

$$\int_{-\infty}^{\mu} d\sigma(\lambda)\,.$$

When $n = 1$, it is exactly solved. As for the matrix case, the set of matrices is not totally ordered. So it is unknown whether such a problem is solvable and how it can be set at all.

It turns out that there exists another representation of an entire operator $\widehat{A}$. The canonical differential equation

$$J \frac{d\varphi}{dt} = \lambda H(t)\varphi, \quad t \in [0, l], \qquad (3.11)$$

is associated with it. Here

$$H(t) = \left(\begin{array}{cc} H_{11}(t) & H_{12}(t) \\ H_{21}(t) & H_{22}(t) \end{array} \right)$$

is a certain function taking its values in the set of $(2n \times 2n)$ Hermitian matrices, acting in the direct orthogonal sum $M \oplus M$, hence, $H_{ik}(t)$ are the n-dimensional blocks,

$$J = \begin{pmatrix} 0 & I_M \\ -I_M & 0 \end{pmatrix}.$$

The set of all functions $\sigma(\lambda)$ coincides with the set of all spectral functions of a certain boundary value problem for the equation (3.11), not complete in the sense that the boundary condition is given only at the left end. The function $f_u(z)$ in (3.8) can be interpreted as the generalized Fourier transform of the corresponding $2n$-tuple $f_H(t)$ from the space of vector-functions with the property

$$\int_0^l f_H(t)H(t)f_H^*(t)\,dt < \infty,$$

by using the normed fundamental solution of the equation (3.11).

We have just considered the "deficient" operators i.e., the operators whose defect numbers are not equal to zero. This fact was emphasized by the hat-sign. Let now A be a self-adjoint operator on $\mathfrak{H}$ and

$$A = \int_{-\infty}^{\infty} \lambda\,dE_\lambda$$

its spectral expansion. Take a subspace $M \subset \mathfrak{H}$. We can regard it as a minimal generating one, that is

$$\overline{\operatorname{span}\left\{E_\lambda u\right\}_{u\in M}} = \mathfrak{H}.$$

We shall call an orthogonal projector P on $\mathfrak{H}$ an entire projector of the operator A if

$$P\mathfrak{H} \supset M$$

and the restriction $\widehat{A}_P$ of A to $P\mathfrak{H}$

$$\widehat{A}_P f = Af,$$
$$\mathcal{D}(\widehat{A}_P) = \{f \in \mathcal{D}(A) \bigcap P\mathfrak{H} | Af \in P\mathfrak{H}\}$$

is M-entire. We say that the family

$$\mathfrak{P} = \{P\}$$

of M-entire projectors of A forms a chain if it admits natural ordering, i.e for any two projectors from $\mathfrak{P}$ the range of one of them contains that of the other. The chain is called maximal if it is not a proper part of any other chain of A.

It appears that for each self-adjoint operator with a generating subspace M there exists a maximal chain $\mathfrak{P}$ of M-entire projectors. Moreover, for every entire projector P, one can always find a chain containing this projector. The chain possesses the following fundamental property: for any two operators $P_1, P_2 \in \mathfrak{P}$, either there is an operator $P \in \mathfrak{P}$ between P_1 and P_2 or there is no such operator P. The latter means a break in the chain. In this case the subspace $(P_1 - P_2)\mathfrak{H}$ is one-dimensional. The operator A can be considered as a limit of the extending system of the M-entire operators $\widehat{A}_P$.

Having fixed a chain $\mathfrak{P}$, it is possible to introduce the notion of a compactly supported vector.

We call a vector $f \in \mathfrak{H}$ compactly supported if

$$f \in P\mathfrak{H}$$

for some $P \in \mathfrak{P}$.

If a vector $f \in \mathfrak{H}$ is compactly supported, then it is represented by the entire function $f_M(z)$ taking its values in the subspace M. This representation does not depend on the choice of $P \in \mathfrak{P}$. So, we have a dense set of elements representable by the entire functions. In addition, the operator A is transformed into the multiplication by z.

On the other hand, there exists another representation of the element f. The canonical equation of the form (3.11), but now on the whole half-axis, is associated with A. The mapping

$$f \mapsto f_H(t), \qquad (3.12)$$

where $f_H(t)$ is related to $f_M(z)$ by the equality

$$f_u(\lambda) = \int_0^\infty f_H(t)\varphi(t, \lambda)H(t)\, dt,$$

$\varphi(t, \lambda)$ is the fundamental solution of the canonical equation, is isometric in the sense that

$$(f, f) = \int_0^\infty f_H(t)H(t)f_M^*(t)\, dt.$$

But what are the advantages of the last representation? The principal one is that the operator A represented by the multiplication by λ in the spectral representation is transformed under the map (3.12) into the operator

$$H^{-1}(t)J\frac{d}{dt}.$$

I would like to note that I started to develop the theory of entire operators in 1942. My papers of 1942–1943 concerning one important class of operators were devoted mainly to entire Hermitian operators whose defect numbers equal 1. In 1949 I constructed the theory of entire operators with an arbitrary finite deficiency index (n, n). But at that time I did not yet understand the relation of this theory to the triangular representations of self-adjoint operators. Frankly speaking, I could not comprehend this because the remarkable works by V. Potapov and M. Livshitz on analytic matrix-functions had not yet been written. Their investigations considerably simplify the techniques of representing such operators. The latest results by M. Brodsky, V. Matsaev, I. Gohberg, and mine on the theory of non-self-adjoint operators enable us to get a lot from the case of infinite defect numbers.

I would like also to draw attention to the difference between the problem under discussion and the inverse problems considered by I. Gelfand, B. Levitan, V. Marchenko and myself. In these authors' investigations a differential operator is given in advance, and then it is clarified how this operator can be regenerated by its spectral function and which properties the spectral function must have to make it possible. When we deal with the above situation, we have no preassigned differential operator at all. We do not even know in advance the form of such an operator. There is only a self-adjoint operator, but the corresponding differential equation is determined by the statement of the problem which is quite general. This differential operator was found to be

$$H^{-1}(t) J \frac{d}{dt} .$$

This is its most general form, servicing all self-adjoint operators. The evolution of its writing was complicated. In the very beginning, it was required that the coefficients of the determined self-adjoint differential operator be smooth enough, then the prefix "quasi" was added to the word "differential", the Stieltjes differentials appeared, and so on. But as we could see, all these operators might be reduced to the form $H^{-1}(t) J \frac{d}{dt}$, and there is no more general one. This is the relation of the problem under discussion to V. Potapov's results revealed by me in 1957. There are the notes by one of my students of all these results concerning the case of the deficiency index (n, n), $n < \infty$. But I did not publish them. I deemed it necessary to have the uniqueness theorem. I was looking for the conditions under which the chain of the entire projectors P of the operator A is unique. In general, it is not unique. But I have shown that if $n = 1$ and A is real in a certain involution, then in many cases this chain and, hence, the function $H(t)$, is uniquely determined. I assumed that this was always true (for $n = 1$), provided that the normalization condition sp $H(t) = 1$ is fulfilled.

However, shortly before I was honoured to give the lecture here, I discovered unexpectedly, that my hypothesis for $n = 1$ had been verified by the American mathematician L. de Branges. Of course, his result is not formulated in operator theory terms. He considered a two-dimensional Hamiltonian $H(t)$ with the real coefficients, normalized in just the same way as I did, and proved that $H(t)$ is uniquely determined by the spectral function of the canonical equation. Thus, in the case of $n = 1$ the problem has been completed. If the operator is positive, then the certain string is uniquely associated with it. In any other case, the corresponding operator $H^{-1}(t)J\frac{d}{dt}$ is generated by the Hamiltonian $H(t)$.

Some words about my extraordinary and accidental acquaintance with the works by L. de Branges. One of my collaborators informed me of the problems raised in L. de Branges' talk "New and old problems for entire functions" at the meeting of the American Mathematical Society (Madison, 1963). These were the problems related to N. Akhiezer's, S. Bernshtein's, B. Levin's, and S. Mergelyan's investigations. One of the chapters of Levin's monograph published in English and German is devoted to some of these questions. That was what had stimulated my interest in his talk. When studying it in detail, I saw that he introduced certain Hilbert spaces of entire functions there. Then I found the series of his papers related to these spaces under the title "Hilbert spaces of entire functions" where the entire scalar operators ($n = 1$) introduced by me as far back as in 1942 were considered. I find his work to be brilliant. In a short period of time he has covered the distance which took me too many years. However, at the time I was dealing with those problems, a lot of results needed for the construction of the theory of entire operators had not been discovered yet. I had to prove some of them myself. Today, one can find them in the books by R. Boas or B. Levin. Referring, as a rule, to Boas, L. de Branges repeated many of my statements. But the final result concerning not only positive but any entire operator with deficiency index (1,1) was established by him. This is what I didn't have. I was aspiring to it, but had not got there.

Shortage of time prevents me from speaking of, how for instance, extrapolation problems for stochastic processes in the complex plane are solved. I shall restrict myself to a few general comments.

In the fall of 1940 I delivered a talk on the continuation problem for Hermitian-positive functions here. After the talk A. Kolmogorov called my attention to the fact that it was possible to make geometrical sense of this problem, and that it was connected with the theory of spiral lines in a Hilbert space. I did not immediately understand this relation. But in 1943 I already talked at the Society on the continuation theory for spiral lines constructed by me in the war days. During the talk, Kolmogorov remarked that a spiral line was not completely determined even if its whole "tail" from $-\infty$ to the point t was given. This remark seemed strange to me. But I soon found the criterion for a spiral line, i.e. for a continuous process with stationary stochastic differences, to admit an unbounded extrapolation. The result was reduced to Kolmogorov's corresponding one concerning discrete stationary processes. Then I talked on spiral lines in the Lobachevsky space. Later, I found that all these issues are of great importance for the inverse problems of the spectral analysis of differential operators. I reported the results on a string and then on general canonical systems. But these were separate parts of my investigations. So, when I was invited to deliver a lecture at this meeting, I decided to present the survey of all these topics here. If we look at their history, we shall find that the theory of entire operators goes back to the remarkable investigations by P. Chebyshev, A. Markov, and T. Stieltjes on the moment problem. Their main tools were continued fractions. But they led to an impasse. The way out of the situation was found by H. Hamburger, R. Nevanlinna, M. Riesz, E. Hellinger, and T. Carleman. While H. Hamburger was still broadly using continued fractions, the others developed and applied different methods. At the same time, the important relations to various parts of analysis were observed. For example, Nevanlinna discovered an essential connection with extrapolation problems for functions in the classical analysis. The deadlock seemed to come again. But new continual problems appeared, in particular the continuation problems for Hermitian-positive functions, spiral lines, etc. Now, we have come to the problem of representation for Hermitian operators. I regard this problem as being at its initial stage.

I have considered the case of the deficiency index (n, n), $n < \infty$. But it is necessary to study the case $n = \infty$, because it is the very case that covers partial differential equations. If we have the latter completely solved, we shall have known the most general form in which, for instance, an arbitrary elliptic operator can be written, which means finding any self-adjoint extension of the operator Δ, and so on. The hardest point is the uniqueness problem. Even in the case $n < \infty$, $n \neq 1$, this has not been solved.

Comments

Chapter 1

Section 1. The presentation (without proof) of the theory of Hermitian operators in a Hilbert space follows the book by N.I. Ahiezer and I.M. Glazman [1] which contains complete information on the topics considered in the section.

Section 2. The foundations of the representation theory of entire operators are shown there. All the results, except for Theorem 2.4, belong to Krein. They are announced in his papers [2,3]. The detailed proof can be found in [10]. As for Theorem 2.4, it is due to M.S. Livshitz. It was established in his D.Sc. dissertation [1].

Section 3. Krein's results on the structure of a spectrum of self-adjoint extensions within the original space of a Hermitian operator are given on the basis of his article [7].

Section 4. The theory of subharmonic functions is presented following the monographs by A.I. Markushevich [1] and I.I. Privalov [1]. The class of N-functions was introduced by A. Ostrovsky and the brothers F. and R. Nevanlinna. Theorems 4.1 and 4.2 clarifying some important properties of N-functions were obtained by R. Nevanlinna [1] and V.I. Smirnov respectively [1] (one can find the proof in the book by I.I. Privalov [2]). Lemmas 4.1–4.3 and Theorems 4.3–4.5 characterizing the growth of N-functions were proved by Krein. They are given in accordance with his paper [8]. Lemmas 4.4, 4.5 are also due to Krein [10]. The relation of the indicator of an entire function of exponential type to the support function of a certain convex set (the Polya theorem, see G. Polya [1]) and that of its zeroes to the growth at the infinity are described as in B.Ya. Levin's monograph [1].

Chapter 2

Sections 1–6. The results are devoted to the theory of entire operators and connected problems. They belong to Krein [2,3,6] and [10]. The exception is subsection 6.1 where the well-known Riesz-Herglotz theorem and statements concerning the properties of R-functions are expounded in the style of the book by N.I. Ahiezer and I.M. Glazman [1].

Section 7. The description of all distribution functions of an entire Hermitian operator is due Krein. The results were announced in his work [3]. The details are taken from the notes of his unpublished lectures. The relation of entire operators to canonical systems of differential equations was well understood by Krein after

V.P. Potapov's paper [1] appeared. But still the question of one-to-one-ness of the correspondence between them had remained open. In Krein's lectures it is formulated as an almost undoubted hypothesis under a certain normalization condition on the Hamiltonian of the differential system. The proof was obtained later by L. de Branges [1]. One can find the historic discussion in Appendix 3.

Section 8. The method of directing functionals was discovered by Krein and published in his article [9]. The entire operators which have in addition directing functionals are investigated in [10].

Section 9. Krein pointed out repeatedly the necessity of studying Hermitian operators entire with respect to a generalized gauge (see, for instance, his talk [11] at ICM-66). The theory of such operators is presented following the works by Yu.L. Shmulian [1–6] the complete survey of which is contained in paper by E.P. Tzekanovsky and Yu.L. Shmulian [1].

Chapter 3

Section 1. The classical power moment problem is interpreted from the point of view of the theory of entire operators. It should be noted that the expositions of this problem in the books by N.I. Akhiezer [1] and Yu.M. Berezansky [1] use approaches different from the one suggested here.

Section 2. The results concerning the continuation problem for positive definite functions belong to Krein [1]. Their presentation, modified in order to apply the theory of entire operators, is given in accordance with the notes of his lectures. The examples of both definite and indefinite cases of the continuation problem for positive definite functions are taken from one of his unpublished manuscripts.

Section 3. A number of concrete problems of extrapolation and filtration of stationary stochastic processes leads to the continuation problem for spiral arcs. In due course A.N. Kolmogorov remarked on the importance of the latter. The complete solution of the problem for spiral and unitary arcs was given by Krein on the basis of the theory of entire operators. The results were partly announced in his papers [4,5]. Their exposition follows the notes of Krein's lectures referred to above.

Appendix 1

The results concerning the integration of operator-functions with respect to an operator measure are due to H. Langer [1] and Yu.L. Shmulian [5,6]. The theory of Hermitian operators with arbitrary defect numbers was developed by Krein and Sh.N. Saakyan and is presented in the scheme of their joint papers [1,2]. It is to be noted that the functions of the first and the second kind in the more general situation of a Banach space were introduced and studied by I.Ts. Gohberg and

A.S. Markus [1]. The theory of representations of operators entire with respect to a generalized gauge whose deficiency index is arbitrary is exposed following Yu.L. Shmulian's works [1–6]. In these papers, the problem appearing in the Preface as problem 1) set up by Krein at his lecture at ICM-66 (see [11]) is solved.

Appendix 2

Papers by authors [1–3] form the basis of the Appendix which is devoted to solving problem 2) in the Preface put by Krein. The case where an unbounded operator coefficient in the operator-differential expression is missing was investigated in an earlier M.L. Gorbachuk article [1]. If the original space is one-dimensional, i.e. the equation under consideration becomes the ordinary Sturm-Liouville one, the description of all spectral functions in the most general situation was done by I.S. Katz and Krein [1]. The generalization of the notion of a directing functional from subsection 2.5 and Theorem 2.4 are due to H. Langer [1]

Appendix 3

This Appendix is the translation from Russian of the notes of Krein's talk at the Jubilee Session of the Moscow Mathematical Society (1964). The notes were taken from Krein's archive.

Bibliography

Akhiezer N. I.

  1 *Classical moment problem* (in Russian), Moscow, Fizmatgiz, 1961.

  2 *Lectures on approximation theory* (in Russian), Moscow, Nauka, 1965.

Ahiezer N. I., Glazman I. M.

  1 *Theorie der linearen Operatoren im Hilbert-Raum* (transl. from the Russian). Akad. Verlag, 1954.

Berezansky Yu. M.

  1 *Expansion in eigenfuctions of self-adjoint operators* (transl. from the Russian). Amer. Math. Soc., 1968.

Branges L.

  1 *Some Hilbert spaces of entire functions IV*, Trans. Amer. Math. Soc. 105 (1962), 43–83.

Danford N. and Schvartz J. T.

  1 *Linear operators. General theory 1.*, New York, London, Interscience Publishers, 1958.

Derkach V. A. and Malamud M. M.

  1 *Generalized resolvents and boundary value problems for Hermitian operators with gaps*, J. of Functional analysis 95 (1991), No. 1, 1–95.

Gel'fand I. M., Shilov G. E.

  1 *Generalized functions Vol. 2*, Acad. Press, 1968 (transl. from the Russian).

Gohberg I. Ts. and Krein M. G.

  1 *The theory of Volterra operators in a Hilbert space and applications* (in Russian), Moscow, Nauka, 1967.

Gohberg I. Ts. and Markus A. S.

  1 *Characteristic properties of some points of spectrum of linear bounded operators* (in Russian), Izvestiya Vysch. Uchebn. Zavedeniy, Matematika, 2 (15) (1960), No. 5, 74–87.

Gorbachuk M. L.

  1 *On spectral functions of a second order differential equation with operator coefficients* (in Russian), Ukrain. Mat. J. 18 (1966), No. 2, 3–21.

Gorbachuk V. I., Gorbachuk M. L.

1 *Expansion in eigenfunctions of a second-order differential equation with operator coefficients,* Dokl. Akad. Nauk SSSR 184 (1969), No. 4, 774–777. (Transl.: Soviet Math. Dokl. 10 (1969), 158–162.)

2 *An example of an entire operator whose defect numbers are infinite* (in Ukrainian), Dokl. Akad. Nauk Ukr. SSR. Ser A. No. 7 (1970), 579–582.

3 *Problems of the spectral theory of a second-order linear differential equation with unbounded operator coefficients,* Ukrain. Mat. Zh. 23 (1971), No.1, 3–15. (Transl.: Ukr. Math. J. 23 (1971), 1–12.)

Hardy G., Littlewood J., Polya G.

1 *Inequalities* (Russian transl.), Moscow, IL, 1948.

Katz I. S. and Krein M. G.

1 *On spectral functions of a string.* Appendix 2 to the book "Discrete and continuous boundary problems" by F.V. Atkinson (Russian transl.), Moscow, Mir, 1968.

Krein M. G.

1 *On continuation problem for positive definite functions* (in Russian), Dokl. Akad. Nauk SSSR 26 (1940), No. 1, 17–21.

2 *On Hermitian operators with defect numbers one* (in Russian), Dokl. Akad. Nauk SSSR 43 (1944), No. 8, 339–342; 44 (1944), No. 4, 143–146.

3 *One remarkable class of Hermitian operators* (in Russian), Dokl. Akad. Nauk SSSR 44 (1944), No. 5, 191–195.

4 *On continuation problem for spiral arcs in a Hilbert space* (in Russian), Dokl. Akad. Nauk SSSR 45 (1944), No. 4, 147–150.

5 *On the A. N. Kolmogoroff extrapolation problem* (in Russian), Dokl. Akad. Nauk SSSR 46 (1945), No. 8, 339–342.

6 *On resolvents of a Hermitian operator with deficiency index (m,m)* (in Russian), Dokl. Akad. Nauk SSSR 52 (1946), No. 8, 657–660.

7 *Theory of self-adjoint extensions of semibounded operators and its applications* (in Russian), Mat. Sbornik 20 (1947), No. 3, 431–495.

8 *Theory of entire functions of exponential type* (in Russian), Izv. Akad. Nauk SSSR, Ser. Mat. 11 (1947), No. 4, 309–326.

9 *On Hermitian operators with directing functionals* (in Ukrainian), Zbirnyk pratz Institutu Mat. Akad. Nauk Ukrain. SSR, 10 (1948), 83–106.

10 *The principal aspects of the theory of a representation of Hermitian operators whose deficiency index is (m,m)* (in Russian), Ukrain. Mat. Zh. (1949), No. 2, 3–66.

11 *Analytical problems and results of the theory of linear operators in a Hilbert space* (in Russian), Proceedings of ICM, Moscow, Mir, 1968, 189–216.

Krein M. G. and Saakyan Sh. N.

1 *Some new results in the theory of resolvents of Hermitian operators* (in Russian), Dokl. Akad. Nauk SSSR 169 (1966), No. 6, 1269–1272 (Transl.: Soviet Math. Dokl. 7 (1966)).

2 *Resolvent matrix of a Hermitian operator and characteristic functions connected with it* (in Russian), Functsion. Anal. i Prilozh. 4 (1970), No. 3, 103–104 (Transl.: Functional Anal. Appl. 4 (1970)).

Langer H.

1 *Über die Methode der richtenden Functionalen von M. G. Krein*, Acta Math. Hungarica 21 (1970), 207–225.

Levin B. Ya.

1 *Distribution of zeroes of entire functions* (in Russian), Moscow-Leningrad, Tekhnik-Teor. Literatura, 1956 (Transl.: American Mathematical Society, Providence, 1964).

Livshitz M. S.

1 *New applications of the theory of Hermitian operators* (in Russian), PhD dissertation, Maikop, 1942.

Markushevich A. I.

1 *Theory of functions of a complex variable* (in Russian), Moscow-Leningrad, Tekhnik-Teor. Literatura, 1950.

Nevanlinna R.

1 *Single-valued analytic functions* (Russian transl.), Moscow, 1941.

Polya G.

1 *Untersuchungen über Lücken und Singularitäten von Potenzreihen*, Math. Z. 29 (1929), 549–640.

Potapov V. P.

1 *Multiplicative structure of J-contractive matrix-functions* (in Russian), Trudy Moskov. Mat. Obšč 4 (1955), 125–136.

Privalov I. I.

1 *Subharmonic functions* (in Russian), Moscow-Leningrad, Tekhnik-Teor. Literatura, 1937.

2 *Boundary properties of single-valued analytic functions* (in Russian), Moscow-Leningrad, Gostekhizdat, 1950.

3 *Introduction to the theory of functions of a complex variable* (in Russian), Moscow-Leningrad, Fizmatgiz, 1960.

Schwartz L.

1 *Théorie des distributions à valeur vectorièlles*, Ann. Inst. Fourier, 7 (1957), 1–141.

Shmulian Yu. L.

1 *Extended resolvents and extended spectral functions of a Hermitian operator* (in Ukrainian), Dokl. Akad. Nauk Ukr.SSR. Ser.A. (1970), No. 3, 230–234.

2 *Representation of a Hermitian operator with a generalized module* (in Ukrainian), Dokl. Akad. Nauk Ukr.SSR. Ser.A. (1970), No. 5, 432–435.

3 *Direct and inverse problem of the resolvent matrix theory* (in Ukrainian), Dokl. Akad. Nauk Ukr.SSR. Ser.A. (1970), No. 6, 514–517.

4 *Extended resolvents and extended spectral functions of a Hermitian operator* (in Russian), Mat. Sbornik 84 (1971), No. 3, 440–455.

5 *Representation of Hermitian operators with a generalized gauge* (in Russian), Mat. Sbornik 85 (1971), No. 4, 553–562.

6 *Operator nodes* (in Russian), Mat. Issledovaniya 8 (1973), No. 2, 147–160.

Smirnov V. I.

1 *Sur les valeurs limites des functions regulières a l'intérieur d'un cercle*, Zh. Leningradskogo Mat. Obšč 2:2 (1928), 22–27.

Tzekanovsky E. P. and Shmulian Yu. L.

1 *Aspects of the extension theory for unbounded operators in rigged Hilbert spaces* (in Russian), Itogi nauki i tekhn. Mat. Analiz 14 (1977), 59–100.

Subject Index

Titles previously published in the series

OPERATOR THEORY: ADVANCES AND APPLICATIONS
BIRKHÄUSER VERLAG

Edited by
I. Gohberg,
School of Mathematical Sciences, Tel-Aviv University, Ramat Aviv, Israel

This series is devoted to the publication of current research in operator theory, with particular emphasis on applications to classical analysis and the theory of integral equations, as well as to numerical analysis, mathematical physics and mathematical methods in electrical engineering.

78. **M. Demuth, B.-W. Schulze** (Eds): Partial Differential Operators and Mathematical Physics: International Conference in Holzhau (Germany), July 3–9, 1994, 1995, (ISBN 3-7643-5208-6)

79. **I. Gohberg, M.A. Kaashoek, F. van Schagen**: Partially Specified Matrices and Operators: Classification, Completion, Applications, 1995, (ISBN 3-7643-5259-0)

80. **I. Gohberg, H. Langer** (Eds): Operator Theory and Boundary Eigenvalue Problems. International Workshop in Vienna, July 27–30, 1993, 1995, (ISBN 3-7643-5275-2)

81. **H. Upmeier**: Toeplitz Operators and Index Theory in Several Complex Variables, 1996, (ISBN 3-7643-5282-5)

82. **T. Constantinescu**: Schur Parameters, Factorization and Dilation Problems, 1996, (ISBN 3-7643-5285-X)

83. **A.B. Antonevich**: Linear Functional Equations. Operator Approach, 1995, (ISBN 3-7643-2931-9)

84. **L.A. Sakhnovich**: Integral Equations with Difference Kernels on Finite Intervals, 1996, (ISBN 3-7643-5267-1)

85/ **Y.M. Berezansky, G.F. Us, Z.G. Sheftel**: Functional Analysis, Vol. I + Vol. II, 1996,
86. Vol. I (ISBN 3-7643-5344-9), Vol. II (3-7643-5345-7)

87. **I. Gohberg, P. Lancaster, P.N. Shivakumar** (Eds): Recent Developments in Operator Theory and Its Applications. International Conference in Winnipeg, October 2–6, 1994, 1996, (ISBN 3-7643-5414-5)

88. **J. van Neerven** (Ed.): The Asymptotic Behaviour of Semigroups of Linear Operators, 1996, (ISBN 3-7643-5455-0)

89. **Y. Egorov, V. Kondratiev**: On Spectral Theory of Elliptic Operators, 1996, (ISBN 3-7643-5390-2)

90. **A. Böttcher, I. Gohberg** (Eds): Singular Integral Operators and Related Topics. Joint German-Israeli Workshop, Tel Aviv, March 1–10, 1995, 1996, (ISBN 3-7643-5466-6)

91. **A.L. Skubachevskii**: Elliptic Functional Differential Equations and Applications, 1997, (ISBN 3-7643-5404-6)

92. **A.Ya. Shklyar**: Complete Second Order Linear Differential Equations in Hilbert Spaces, 1997, (ISBN 3-7643-5377-5)

93. **Y. Egorov, B.-W. Schulze**: Pseudo-Differential Operators, Singularities, Applications, 1997, (ISBN 3-7643-5484-4)

94. **M.I. Kadets, V.M. Kadets**: Series in Banach Spaces. Conditional and Unconditional Convergence, 1997, (ISBN 3-7643-5401-1)

95. **H. Dym, V. Katsnelson, B. Fritzsche, B. Kirstein** (Eds): Topics in Interpolation Theory, 1997, (ISBN 3-7643-5723-1)

96. **D. Alpay, A. Dijksma, H. de Snoo**: Schur Functions, Operator Colligations, and Reproducing Kernel Pontryagin Spaces, (ISBN 3-7643-5763-0)